Indigenous Industries of Assam

Indigenous Industries of Assam

Retrospect and Prospect

PRIYAM GOSWAMI

Maulana Abul Kalam Azad Institute of Asian Studies, Kolkata

ANSHAH Anshah Publishing House

Published on behalf of Maulana Abul Kalam Azad Institute of Asian Studies, Kolkata. Views expressed are those of the author/contributors and do not reflect the position of the Institute and s/he are solely responsible for any copyright violation.
(First Published in India in 2005)

Rs. 995

ISBN: 978-8-18-364005-3

2027
PoD Impression

Indigenous Industries of Assam: Retrospect and Prospect

Published by:
ANSHAH PUBLISHING HOUSE
LG 18-19, Pankaj Central Market,
IP Ext., Patparganj, Delhi 110092
11 47322068; 9650028065, 9810522367
info@shiprapublication.com
www.shiprapublication.com

Preface

This book is the result of my investigation which I pursued with financial assistance from Maulana Abul Kalam Azad Institute of Asian Studies, Kolkata. In course of my earlier study on colonial penetration in Assam, and thereafter, the necessity for a study of the inadequate industrial development of Assam has always been at the back of my mind. In any agricultural economy, progress in the agricultural sector must be simultaneous with industrial growth if an overall economic development is to be expected. This can be best achieved if industrialisation in a particular region is based on the full utilisation of the industrial skills of the indigenous population rather than the superimposition of an imported technology unfamiliar to them. Rural industries form the backbone of a developing country. In a state like Assam, where more than 85 per cent of the population live in villages, the promotion of this sector is of utmost importance. Moreover, in the context of large scale unemployment among youth, industrialisation is the only way through which avenues for sustainable and productive employment can be generated. But development plans for industry cannot be framed in isolation. They must be integrated with plans for development in all other sectors — agriculture, transport, power — in order to ensure success. While pursuing the present study, efforts were made to test these hypotheses with the help of empirical data.

This work would not have been possible had it not been for the help that I received from several quarters. I am extremely grateful to the Maulana Abul Kalam Azad Institute of Asian Studies for its generous support. Professor Mahavir Singh, the dynamic Director of the Institute has been a source of constant encouragement and I am grateful to him for his help. My thanks are also due to my friends and colleagues with whom I shared many of my ideas and whose suggestions helped me to formulate my thoughts. Finally, I would like to acknowledge with gratitude the help that I received from the numerous artisans with whom I interacted during the course of my study.

Priyam Goswami

Contents

SECTION II
Prospects of Indigenous Industries

INTRODUCTION

Medieval Assam—Socio-economic Structure

Medieval Assam, relatively isolated from the rest of India, had a unique socio–economic structure. The simple village economy was more or less self-sustaining though not entirely self-sufficient. Apart from salt, which was imported in considerable quantities from Bengal,[1] most other requirements were available locally. Agriculture was the mainstay of the people and communal ownership of land was a basic feature of this agricultural economy. Even within the Ahom feudal framework, where one's position in society was largely determined by birth, the king, the nobility, the priests and other vassal chiefs, all had agricultural lands cultivated by *paiks*.[2] The *paiks* themselves were primarily peasant cultivators whose hereditary private proprietary rights existed only in the case of homesteads and gardens but not in the case of paddy lands. Each *paik* received two *puras*[3] of wet paddy land as *ga mati* in lieu of which he had to render three or four months of service to the State. In addition to his ancestral homestead lands, a *paik* could obtain inferior land if he desired. Additional wet paddy lands were also allotted in certain cases, but only if surplus land (*ubar mati*) was available. The king, as the representative of the community, generally established his claim to the ownership of all communal wet paddy lands and waste lands. It appears that nominal taxation on additional holdings was introduced, though only during the later part of Ahom rule. The entire taxed portion of land or a portion of *ga mati* could be taken back by the Crown if necessary. Thus, the basic pattern of the land system and distribution rested on the concept of communal ownership of land and land revenue consisted in the personal labour service of the peasant *paik*.[4]

Another significant feature of Assamese rural society was the fact that although primarily agriculturists, the people combined agriculture with other trades. There are many references to weavers, spinners, blacksmiths, goldsmiths, potters and workers in ivory, wood, dyes, hide and cane. Almost all the people were self-employed and manufactured their products in their own homes, generally for their own use. There was little specialisation and indigenous crafts and industries formed an integral part of every household. Spinning and weaving, for example, were a part of every woman's work irrespective of caste or status. Similarly, the extraction of mustard oil or the production of jaggery (*gur*) from sugar cane was generally carried out in the individual households. The little specialisation that did exist was in the production of such items as bell metal utensils, pottery or gold jewellery. In contrast, the production of guns and cannon balls, an industry controlled by the State, had acquired a high level of perfection.[5] Generally each village was a self-sustaining unit and its autonomy was a significant feature of medieval Assamese society. Further, unlike in many other parts of India no worker in Assam was hereditarily attached to any particular trade in such a way as to tie him down to a particular community.[6] Hence there was considerable flexibility in the social structure so far as the practice of trades was concerned.

Within this overall framework of a subsistence economy where surplus was limited, the scope for trade and commerce was naturally restricted. In the absence of a monetary economy, moreover, barter was the prevailing mode of transaction.[7] Any discussion on the economy of Assam, however, has to take into consideration certain important factors like the geographical isolation of the region, the isolationist tendencies of her rulers and the consequent insular attitude of the people which together contributed to the unique nature of the Ahom economy. The Ahoms ruled Assam for almost six centuries. During this period they produced shrewd statesmen whose actions were determined not by sentiment but by political considerations and practical expediency. Therefore, as in their social and religious outlook, the Ahom rulers were equally adept in their commercial policy. Although basically isolationists, they encouraged trade to a limited extent[8] so that the people of the neighbouring territories could frequent them with commodities and thereby contribute towards the maintenance of steady income from market duties and customs revenue. This trade, which was accomplished through an organised market system in the adjoining foothills and plains, passed through the *Duars* or passes in the North Eastern and North Western borders of the Brahmaputra Valley. For the Ahom Government, however, the frontier trade, which formed the basis of the plains-hills contact, meant much more than a source of revenue. It was a symbol of assertion of their power beyond the physical limits of their kingdom. Trade with Bengal was limited to bartering a single item of import, salt, for a range of assorted products from Assam.[9]

What is of significance is the fact that apart from rice, mustard seeds and other agricultural products, important items of export from Assam were silk cloths, lac, dyes, ivory, iron hoes, bell metal vessels and cotton broadcloth—all manufactured items.[10] It is thus apparent that within the limited parameters, manufactured items from Assam were in considerable demand in the neighbouring regions and formed an important component of the trade. Thus, it cannot be denied that the indigenous crafts and industries were a distinctive and integral feature of the economy of Assam. The legacy of this rich tradition is evident even today.

British Assam—Economic Potential

When the British occupied Assam after the first Anglo-Burmese War (1824-26) and the subsequent Treaty of Yandabo, they found that years of disruption and turmoil had wrought havoc in the medieval kingdom.[11] The administrative structure was non-existent, the economy was in shambles and the overall picture was one of loot, plunder and decay. The country was in a precariously depopulated state; important administrative and trading centers had been reduced to mere clusters of dilapidated houses, while most of the roads had been overgrown with deep and sometimes, impenetrable jungles. The general destruction was apparent. Yet in spite of such conditions the British decided to stay on. Haunted by fears of a renewed Burmese invasion, it was no doubt strategically important for them to ensure that the region did not relapse into anarchy. But there was another equally important factor which influenced their decision not to vacate the occupied territory immediately. More than two years of stay in the area had convinced them of the economic potential of the region. Taking into consideration the immense economic transformation that had occurred in England following the Industrial Revolution, the East India Company realised that it would indeed be foolish to ignore this fact.

Nineteenth century England had seen a changeover from the phase of merchant capitalism to industrial capitalism where emphasis shifted from revenue collection and trade to new forms of surplus appropriation. As European trade diminished, the vacuum was sought to be filled up by the trade with China, Tibet and Burma. Assam, it was hoped, would serve not only as a market and a forwarding agency, but as a rich hinterland of Calcutta as well. The East India Company quickly set about working towards this end. The expectations were more than fulfilled and within half a century, the entire Brahmaputra Valley underwent a dramatic change. The transition was swift. The changeover from the old order to the new was characterised by a complete overhauling of the administrative machinery which heralded far reaching political, social and economic changes in Assam.

Meanwhile, England had emerged as a major industrial power and India had become the main market for British manufactured goods and an important source for a variety of raw materials. Hence all commercial activities in India were geared to the growing external trade. One particularly important outcome of this policy was that relating to the process of the deindustrialisation of Indian crafts and industries. From the outset, the British aimed at suppressing the local industries and replacing indigenous products with British goods. A significant method of achieving this was the discriminatory customs duties which narrowed the English market for Indian goods through prohibitions and extra-high tariffs. On the other hand, the Indian market was opened up for British manufacturers by the imposition of Free Trade in India.[12]

The general British attitude was reflected by a comment of Doyle and Parkins, two officers of the Department of Customs in India., who once wrote:[13]

> "As in all semi-barbarous countries, the manufacturing industry of India will decline in proportion as its intercourse with a civilised manufacturing country increases, and the attention of the people will rather be turned to the improvement of raw produce exchangeable for the manufactured goods of England."

British Policy and Assamese Industrial Sector

Assam also fell prey to this general British policy. The traditional Assamese crafts and industries sector, which had been sustained for centuries, suddenly found itself in a precarious position. In the face of an onslaught of cheap, machine-made imported items, the technologically backward local industries found themselves in a precarious condition. It became increasingly difficult for the people to hold on to their crafts and within a short period they completely lost the grip on most items. As a result, like in the rest of India, here also the peasants lost the subsidiary income which they had formerly earned when they were able to combine industry with their agricultural pursuits. Consequently, the pressure on land increased.

The significant fact is the ease with which Assam capitulated without any resistance. Edward Gait ascribes this to the absence of occupational sub-castes among the Assamese. He believed that it was because of this that the imported articles could supplant the indigenous crafts so easily.[14] While agreeing that this could have been an important factor, it cannot be accepted as the sole factor. In the case of Assam the additional factors which augmented the deindustrialisation process included the low level of technical skill, the internal disorders prevailing in the kingdom and the Burmese occupation which preceded British occupation. Commenting on the prevailing situation at that time, Lt. Col. Francis Jenkins wrote:[15]

> "When we assumed charge of Assam nothing could possibly be more unpromising than the state of the country. The small remnant of the people had been so harassed and oppressed by long civil and internal wars....that they had given up cultivation, and lived on jungle roots and plants, and famine and pestilence carried off thousands that had escaped the sword and captivity."

In such adverse circumstances, therefore, it could not be expected of the people to put up any significant resistance. In course of time, as the British consolidated their position in the region, the economy of Assam, along with the rest of India, was integrated into the world of modern capitalism though in a subordinate colonial position. The case of Assam was merely an extension of this general process operating in the rest of India. Yet, it is significant that in spite of such conditions, a few industries did survive. A notable example is the handloom industry which was intimately connected with the production of *mekhela chador*, the traditional attire of Assamese women. This industry survived not because it was able to compete with British industrial products, but because in this sector it did not face any competition at all. The dress of the Assamese women, which consisted of *mekhela chador, riha and khonia kapor*, was unique to the region and not initially manufactured outside the province. Moreover, silk competed very largely with cotton in supplying the wearing apparel of the people. Hence the efforts of the British in making inroads in the local handloom industry was considerably delayed. In fact, the silk handloom industry has been able to hold its own against all odds and is a flourishing industry in Assam even today.

Like the silk industry, there are several other industries which are indigenous to this region which, if promoted, can figure not only in the national but in the international market as well. Despite the availability of a wide range of economic resources and a rich tradition of indigenous crafts, the present scenario of cottage and small scale industries in Assam is not at all vibrant. Several factors have contributed to this situation resulting in economic stagnation and poverty. Inadequate infrastructure, lack of sustained Government support, absence of trained manpower and technological skill are important causes, just to name a few.

The Study and Objectives

It is generally agreed that rural industries form the backbone of a developing country. In Assam, 87 per cent of her population live in rural areas (Census of 2001). Hence it is extremely important to mobilise this sector if we want economic progress. Economists postulate a process of development proceeding from an agricultural stage through an agricultural-industrial stage to an agricultural-industrial–commercial stage. A unique feature of the Indian economy since independence has been the rapid growth of the small industries sector which comprises of small scale industries and cottage industries. Small scale industries, generally located in the urban centres as separate establishments, produce goods with partially or wholly mechanised equipment operated by hired labour. Cottage industries, on the other hand, are normally associated with agriculture and provide subsidiary employment to rural areas. They generally involve manual operations and produce goods primarily with the help of the members of the family as a whole or part time occupation.

Assam, with its rich tradition of indigenous crafts and industries and ample economic resources, has immense scope for the growth and sustained development of the small industries sector. However, despite the fact that it can play a pivotal role in the regional economy in terms of employment and economic growth, this potential has not been fully exploited. It is in this context that the present study is being made. The major industries like tea, coal, oil, jute etc do not fall within the purview of this study which has been confined to the plains districts of the Brahmaputra Valley.

The work has been divided into three broad sections: Section I provides a historical background of the indigenous industries of Assam; Section II attempts to analyse the prospects of certain indigenous industries in the state in the present context; Section III is an assessment of the overall situation. The objectives of the study are:

- to provide a historical background of the indigenous crafts and industries of Assam
- to discuss the prospects of the indigenous crafts and industries of Assam in the national and international market
- to analyse the causes of the present stagnation to suggest probable remedies.

References

1. Almost 1,00,000 *maunds* of salt used to be annually imported from Bengal. R.B. Pemberton, *Report on the Eastern Frontier of British India,* p. 82.
2. A *paik* was an enrolled adult subject of the Ahom government having specific duties to render to the State for fixed periods of the year. The officers of different categories were allotted tracts of land and a specified number of *paiks* to work for them.
3. Approximately three acres.
4. Shihabuddin Talish who accompanied Mir Jumla to Assam during his expedition in 1662-63 wrote: 'It is not the custom here to take any land tax from the cultivators, but in every house one man out of three has to render service to the Raja'. Talish, *Fathiya-I-ibria,* tr. J.N. Sarkar, Journal of the Bihar and Orissa Research Society, Vol.1, p. 179.
5. Shihabuddin Talish wrote that the Ahoms "cast excellent matchlocks and bachadar artillery and show great skill in this craft'. *Fathiya-I-ibria,* J.N. Sarkar's translation cited

in J.N. Phukan's unpublished thesis '*The Economic History of Assam Under the Ahoms*", Gauhati university, 1973, p. 122.

6. There are numerous examples in India showing that the basic requirement in most Indian cases of successful entrepreneurship has been membership of a particular caste or community, e.g., the Marwaris, the Seths, the Voras, the Komatis, the Gujrati *banias* etc.
7. Local coinage had been introduced from the sixteenth century but on a very limited scale.
8. The Ahom rulers were wary of foreign merchants and did not allow them to settle in Assam. They feared that, as secret agents of some conspiring state, they might create disruption. The foreign traders therefore had to transact their business hurriedly and return to their own countries. S.K. Bhuyan, *Anglo Assamese Relations, 1771-1826,* p. 50.
9. As a substitute for salt, most Assamese used a preparation of ash procured from the burnt barks of plantain trees.
10. R.B. Pemberton, *op. cit.*
11. Taking advantage of the chaotic political situation in Assam, the Burmese had invaded Assam in 1817 and had unleashed a virtual reign of terror. Villages were plundered and burnt; terrible atrocities were committed on the helpless people who suffered untold misery.
12. It is generally accepted that the *laissez faire* doctrine largely influenced British policy in India in the nineteenth century. D.R. Gadgil writes that the attitude of the state till World War I was one of absolute *laissez faire.* In fact even the deindustrialisation thesis rests on the theory that free trade resulted in the decline of indigenous industries and the consequent poverty in India. Although an important factor in the process, it is perhaps an over simplification of facts to label the British policy for the entire period as one of of pure *laissez faire* or 'non-interventionist'. The nineteenth

century, especially, the latter half, witnessed a considerable shift in Government policy. Whenever the interests of the British industrial and commercial clases were at stake, the Government was quick to ignore the non-intervetionist policies of economic liberalism. *Laissez faire* need not necessarily refer to only the positive restraints in the establishment of industries. The active participation of the Government in providing facilities directly beneficial to the industries may also be considered an infringement of the *laissez faire* policy. For details refer P. Goswami, *Laissez faire in N.E. India in the Nineteenth Century: Myth or Reality,* Proceedings of North East India History Association, 1995.

13. N.K. Sinha, *The Economic History of Bengal, 1793-1848,* Vol 111, p. 13. Remarks on Trevelyan's Report on Inland Customs and Town Duties.
14. E. Gait, *Notes on Some Industries of Assam from 1884-95,* p. 124.
15. A.J. Moffat Mills, *Report on the Province of Assam, 1853,* Appendix B. Jenkins to Mills, 23 May, 1853.

SECTION I

HISTORICAL BACKGROUND OF INDIGENOUS INDUSTRIES OF ASSAM

SILK

One of the most prominent industries prevalent in Assam during the pre-colonial period was silk culture and manufacture. The Ahoms greatly encouraged the industry and grants of land were conferred upon the *jogis* or weavers, who were also exempted from personal labour extracted by the state from other classes.[1] Three varieties of silk worm were reared, viz., the *Eri Polu, Pat Polu* and *Muga Polu.* The *Eri* worms (*Attacus Ricini*) were fed on the castor plant leaves or on the leaves of the two main jungle plants called *Kesaru and Gomari.*[2] These worms, which could be reared indoors throughout the year, yielded a thread which was not reeled but spun by hand. The silk was spun into a coarse thread and was woven into large sheets called *Eria Kapor.* The *Pat* worms (*Bombyx Textor*), which were akin to the silkworm of Europe, were fed on mulberry leaves and reared in considerable quantities. There were two varieties of *Pat* worms, viz., the *Bor Polu* and the *Soru Polu,* the difference between the two being that the latter matured early and produced a fairly white silk while the silk spun from the former worm was yellowish. The scientific name of the *Muga* worm, (*Anthreaca Assamoca)* denotes its peculiar connection with Assam. It was found nowhere else in India except Dehra Dun where it occurred sparingly. The *Muga* moth was generally reared on the *Som* trees (*Machilus Addoratisimma)* in the open air. The silk produced from these worms was considered to be the most valuable and was the prescribed attire of the aristocracy.[3] The production of *Muga* silk was an extremely painstaking task and required tremendous care and attention.

Apart from these three varieties of silkworms, there were a number of wild silkworms like the *Bon Muga* (*Attacus Cynthia*), *Bon Pat* (*Ficus indica*), *Kuthuri or* Assamese *tusser* (*Antherea Paphia*). These were, however, destitute of all commercial value because the cocoons were found in a wild state and not in sufficient quantities to be economically viable. All the three domesticated varieties, however, were extremely durable and much in demand. In fact commenting on the specimens of Assamese silk, the Secretary, Foreign Department, wrote in 1830:[4]

> "Judging myself of these specimens compared with the common Bengal silk in the same stage of manufacture, I would say that the Assamese article is much harder, but at the same time shows much greater lustre than the former.... It strikes me we may get valuable articles of the *Materia Medica* from the Assam hills."

The major portion of the silk fabrics were woven primarily for home consumption but any surplus was traded with Bengal and the neighbouring hill tribes for other goods. Of the three varieties, *Eri* was the cheapest and therefore most commonly used. It was also in considerable demand among the hill tribes who procured them at the annual fairs or at the weekly *hats* (markets). In fact, in the early years of colonial rule many Assamese depended on the sale of their *Eri* fabrics to meet their revenue demands.[5] *Muga* and *Pat,* on the other hand, were finer products and the prescribed attire of the higher classes though even the poor possessed a piece or two to be worn on special occasions.

The commercial potential of *Eri, Pat* and *Muga* had been gauged by the British since the early days of their contact with the medieval kingdom of Assam. Hence soon after the occupation of the region, David Scott, Agent to the Governor General North Eastern Frontier, had written to the Government that he felt that of all the commodities available in Assam, silk was likely to be the most profitable. He reasoned that it was one produce that did not necessarily entail continued European supervision. Moreover, as the Assamese were universally acquainted with the preparation of the article unlike in the other states where production was confined to a particular community, he felt that many of the inhabitants would engage in the preparation of silk if it could be shown to be highly profitable.[6] That apart, the export of raw silk would considerably enhance the revenue of the Company's exchequer.[7] Soon thereafter, Scott established a silk farm at Darrang. He brought special plants from Rangpur to improve production but his experiment was not entirely successful.[8] Subsequently, Lister and Company, a private enterprise, also attempted to develop a farm at North Lakhimpur, but shortage of labour proved to be an insurmountable problem and the company was left with no alternative but to abandon the project.[9] Efforts were made by A.C. Campbell, Deputy Commissioner, Kamrup, to set up a silk farm on a commercial scale, but within a short while disease and pestilence destroyed the entire crop. The persistent efforts of Crowe and Mackenzie, two private entrepreneurs, also met with the same fate.[10] Campbell, in charge of one of the most extensive experiments, was therefore convinced that the only way to successfully develop the industry was to invest a large amount of capital and to work on a system analogous to the indigo advances by inducing the villagers to cultivate small patches so as to avoid the outbreak of diseases in an epidemic form which was an important deterrent to the cultivation of silk in Assam.[11] In 1870-71 a survey was conducted to determine the actual amount of land under *Som* cultivation. All Deputy Commissioners were instructed to submit the reports for their districts to the Commissioner's office. The following table[12] has been prepared based on these reports.

District	*Area (in bighas)*	*Revenue (in Rupees)*	*Number Employed*
Kamrup	4,960	2,480	19,840
Darrang	1,100	548	130
Nagaon	1,826	990	1,869
Sibsagar	47,722	23,858	21,000
Lakhimpur	2,793	1,397	4,000
Total	58,401	29,273	46,839

In the meantime certain developments in Europe resulted in an increasing demand for Indian silk. For long, China and Italy were the two main suppliers of silk to England. As Italy became completely subjected to Napoleon's Continental System, Italian silk was effectively withheld from the British market. As a result, the English manufacturers were compelled to rely heavily on the Chinese suppliers. Hence it became very important to search for an additional source. The general belief began to gain ground that if an alternative source could be found there was no reason why it should not be exploited. The prospect that Assam could "contribute large quantities of an article which is produced with so much ease in the valley of the Brahmaputra" roused the interest of many. It was in this context that Assam silk began to play a significant role in the East India Company's silk trade.

In 1884 Stack wrote:[13]

> "Looking simply at their commercial potentialities, these three species of silkworm (*Eri, Pat* and *Muga*) may at once be reduced to two by striking off the mulberry worm (*Pat*) on account alike of the costliness of the silk, the scantiness of the present supply and the difficulty of extending its cultivation. The two remaining species present a much more hopeful enterprise. The *Eri* is more promising of the two, both because it is cheaper and more abundant, and also because being reared indoors its cultivation does not entail that troublesome necessity of watching by day and night which is imposed upon the *Muga* breeder. The *Muga,* indeed, yields a finer silk, but as it is only in the roughest shape that Asam silk can hope to become an article of demand in the English market, the difference of quality will perhaps prove to be a matter of secondary importance".

H.Z. Darrah was of the opinion that *Eri* silk had the greatest potential and that it "would one day form one of the most important staples of Assam".[14] He further felt that since silk required no skill in handling it was "difficult to imagine a combination of circumstances which promised greater success for any other commodity".[15] Allen remarked[16] that:

> "it would be unreasonable to attach too much importance to the failure that has hitherto attended all attempts made by Europeans to cultivate the silk worm. It is inevitable that the pioneers of any industry should make mistakes, and when the subject matter is a delicate insect, these mistakes are likely to have disastrous effects, but further experiments are sure to disclose some means of overcoming the difficulties that have hitherto led to failure."

The price of *Muga* yarn, on an average, was Rs. 8 per *seer* and that of cloth around Rs. 2 per yard.[17] The major portion of the *Pat* and *Muga* produce was consumed locally and only around 20 per cent of it was exported most of which went to Bombay, Dacca and Calcutta and then to the Persian Gulf via these ports.[18] *Eri,* on the other hand, was generally exported and only around 20 per cent retained for home consumption. This coarse material sold from Rs. 6 to Rs. 12 per *seer. Eri* had already found a ready market in Australia and Europe and for some time trade in it was very brisk.[19]

Contrary to Allen's views, however, the failure of the efforts of the British entrepreneurs in setting up a silk industry on a commercial scale in Assam convinced the colonial government of the necessity of viewing the silk trade from a different perspective. Keeping in tune with the general policy of deindustrialisation, the trade in silk was now conceived of in two forms: First, the export of thread, and secondly, the export of cocoons. Of the two, the latter was preferred because indigenous reeling was considered as "unfit for employment for any purpose" by English manufacturers.[20] Hence it was believed that instead of investing in machinery in Assam for the manufacture of silk pieces, it would be more profitable to export the cocoons. The manufacture of silk plushes and similar fabrics out of waste cocoons imported from China was already a flourishing branch of the manufacturing industry in England. There was no reason why this growing demand could not be met from Assam.

The Trade Report of Assam for 1905–06 shows that altogether 643 *maunds* of silk cocoons were exported from the Brahmaputra Valley. As the export trade in cocoons increased, the production of cloth declined. This is evident from the same Report which does not show any export of silk piece goods. Apparently no further attempts were made by the government to manufacture silk on a commercial scale. Stack reported that the *Eri* cocoon had been successfully reeled in Italy and believed that if the method was applied in Assam, the produce could be sold at a much cheaper rate.[21] Unfortunately, this recommendation was not implemented.

Thus, it can be seen that within half a century of British occupation, the silk industry of Assam suffered a severe setback. The reasons were apparent. In the first place, the major portion of the available raw silk was diverted out of the state. Secondly, although silk was locally produced and worn to a certain extent, it was gradually replaced by the imported fabrics[22] which were easier to obtain. Allen commented that "in spite of the high price at present obtainable for silk thread, the villagers find that it pays them better to grow produce for the market than to rear broods of silkworm".[23] Hence, in the face of competition, the silk industry remained a purely household industry catering only to the bare minimal requirements of the households. Significantly, however, unlike many other industries, this industry stagnated but did not die out altogether.

References

1. W.W. Hunter, *A Statistical Account of Assam,* Vol. I, p. 415.
2. E. Stack, *Silk in Assam,* 1884, Shillong, 1896.
3. Hunter, *op. cit.,* p. 138.
4. Foreign Political Consultations, May, 1830, No. 52.
5. B.C. Allen, *Monograph on the Silk Cloths of Assam,* p. 9
6. *Foreign Political Consultations,* 7 May, 1830, No. 52.
7. *Foreign Political Consultations,* 29 October, 1830, No. 36.
8. Stack, *op. cit.*
9. *Ibid.*
10. *Ibid.*
11. *Assam Secretariat Records,* (henceforth A.S.R.), File No. 501, Letter No. 1063 dt. 1 Dec., 1871.
12. *A.S.R.*, File No. 501 of 1869-72., Letter No. 447C, 2 Nov., 1871 from D.C. Kamrup, Letter No. 800, 24 Nov. 1871 from D.C. Darrang, Letter No. 323, 8 Jan. 1872 from D.C. Nowgong, Letter No. 1063, 1 Dec., 1871 from D.C. Sibsagar, Letter No. 795, 20 Feb., 1872 from D.C. Lakhimpur.

13. *Ibid.*
14. H.Z. Darrah, *Note on Eri Silk.*
15. *Ibid.*
16. Allen, *op. cit.*
17. *Ibid.*
18. Stack, *op. cit.*
19. G.N. Gupta, *A Survey of the Industries and Resources of Eastern Bengal and Assam,* 1908, p. 26.
20. Stack, *op. cit.*
21. *Ibid.*
22. Like mill made cotton fabrics, Benrasi *endi,* Bhagapuri *bafta,* Bishnupuri *garvashuti* etc.
23. Allen, *op. cit.*

COTTON

Cotton was another important commodity of Assam. Patches of cotton were grown in the plains of Lakhimpur, Sibsagar, Nagaon and Kamrup districts, primarily for domestic consumption. Known as *Gari* cultivation, it was also cultivated extensively on the lower slopes of the Bhutan Hills and the hilly areas of Goalpara district.[1] There were two main varieties of cotton, viz., *Bor Kopah* and *Soru Kopah*. The *Bor Kopah* was the large-balled, high-growing cotton with few seeds, which therefore, could be ginned easily. This variety could be cropped twice a year. The *Soru Kopah,* on the other hand, was the small-balled species which could be cropped only once a year. The cotton was generally cultivated on forest clearings called *jhums*. A jungle clearing was rarely cropped for more than one season and owing to the easy availability of land, the same clearing was not normally resumed for around ten years.[2] Generally, cotton was cultivated together with other crops like *Ahu* rice, mustard maize, chillies, water melon, brinjals etc. according to the convenience of the cultivator. The main reason for associating a second crop was that cotton grew best if shaded during the initial stages.

The entire process of preparing the cotton from plucking to spinning was carried out by those engaged in its cultivation. There was no special caste in the province, like the *Dhunias* of the North Western Province or the

Pinjas of Punjab, whose occupation was to prepare the cotton for spinning. Generally, each household processed only as much cotton as it required. However, a certain quantity of cotton was cultivated for the purpose of trade as well and "boatloads of it could be seen going down the Kopili and Kollong rivers during the cold season".[3]

The situation changed drastically within a few decades of British occupation. The consolidation of British rule in the region was accompanied by control of the market. As elsewhere in India, here also the local markets were flooded with the products of her industries, the process being facilitated with the introduction of steamers on the Brahmaputra since the 1840s[4] and the establishment of railways in 1882.[5] Initially, However, imported fabrics met with considerable resistance in Assam. This was primarily because the dresses of Assamese ladies were unique to the region and were not manufactured outside the province. In any case, weaving in Assam was a part of every woman's household duties and all the requirements of the household were met by them. Moreover, silk competed largely with cotton in supplying the wearing apparel of the people. This apart, it was generally believed that the sizing of English cotton goods and the pressure applied to the cloth made machine-made fabrics less durable than the home-made fabrics. All these factors combined to delay the penetration of imported fabrics in Assam. Even as late as 1897, Samman commented that "it would probably be long before these are altogether superseded by imported fabrics".[6] In course of time, however, stress of competition and the advancement of new ideas slowly broke down barriers and both the Assamese male and female population began to use the cheaper Manchester cloth. As rapid advances were made in improvement of technology, British manufacturers were able to push their products in the local piece-goods market.

The Table 1 shows the import of cotton yarn and qiece-goods, both European and Indian, into the Brahmaputra Valley.[7]

Table 1: Import of Cotton into Assam

Year	*European Yarn (in maunds)*	*Indian Yarn (in cwt)*	*European Piece-goods (in Rupees)*	*Indian Piece-goods (in Rupees)*
1881	3,925	-	43,07,603	45,630
1882	4,668	17	46,63,274	19,100
1883	4,528	28	53,56,010	26,064
1884	4,088	101	47,33,477	68,372
1885	9,467	169	82,81,020	61,040

By 1906-07, the import of yarn into the Brahmaputra Valley had reached 20,485 *maunds.*[8]

The figures reveal the gradual increase in both imported yarn and imported piece-goods. This is significant because cotton continued to be cultivated in Assam in considerable quantities.[9] Moreover, the average produce of cleaned cotton per acre was estimated at 150 lbs. The proportion of cleaned to uncleaned cotton was in the ratio of 2:5[10] which proved that Assam cotton gave both a larger out-turn per acre and a larger proportion of fibre to seed when compared to cotton from other parts of India.[11]

The official position was that the local people slowed down on spinning their own yarn in favour of imported yarn.[12] This argument is not at all tenable because spinning was an age old practice in Assam and the people were unlikely to give it up altogether so easily. Hence we must look elsewhere for the cause. The Table 2 is likely to provide the answer.

Table 2: Export of Raw Cotton from Assam

Year	*Weight (in maunds)*
1881	20,027
1882	18,464
1883	14,199
1884	17,071
1885	20,277
Total	90,038

After making extensive surveys, A.E. Campbell, Deputy Commissioner of Revenue, Nagaon, reported that "the quantity of cotton exported is yearly increasing".[13] The total quantity of raw cotton exported from Assam in 1906-07 was 101,428 *maunds*[14], more than 11,000 *maunds* over the entire quantity exported between 1881 and 1885.

It is thus apparent that large quantities of cotton were taken out of the province, most of which was exported in an un-ginned state to England through Chittagong and Calcutta. One very important reason for the rapid increase in export was that almost all the varieties of cotton in Assam had short, coarse staples which were in great demand in England for blending with wool or *Eri* to make a variety of products like suit and dress material and blankets.[15] While the former were very popular in Europe, the latter was assured of a ready market in Assam, especially among the tea garden labourers. The Company also needed the cotton from Assam to trade with China, though the Chinese market was rather uncertain and varied according to the production of cotton in China.[16]

The increasing demand for raw cotton led to a series of unsuccessful experiments to introduce superior varieties of cotton in Assam. Eventually it was decided that "the wisest course would be to develop and improve our own short stapled hill cotton for which there is a growing demand in Europe and in other parts of India".[17]

As the export of cotton increased, Assam, which was at one time self-sufficient in cotton products, was compelled to resort to using imported yarn even to meet her own requirements.

References

1. For details refer H.Z. Darrah, *Cotton in Assam,* (1885) *A.S.R.,* Agricultural Ledger No. 8 of 1895.
2. Darrah, *op. cit.*
3. *Ibid.*
4. H.A. Antrobus, *A History of the Assam Company,* Edinburgh, 1957, p. 353.
5. The Assam Railways and Trading Company was incorporated in 1881. The first meter gauge locomotive in Assam passed on the section of railway line between Dibrugarh steamer *ghat* and Jaipur in Upper Assam on 1 May, 1882.
6. Gupta, *op. cit.* p. 7
7. Compiled on the basis of information in Darrah, *op. cit.*,
8. *A.S.R..*, Trade Report for the year 1906-07. (Assam)
9. *A.S.R.,* File No. 498, No. 1-21

10. Darrah, *op. cit.*
11. *Ibid.*
12. Gupta, *op. cit.* p. 3.
13. *A.S.R.*, File No. 498, Letter No. 618, 7 March, 1864.
14. *A.S.R.*, Trade Report for 1906-07 (Assam)
15. Gupta, *Op. cit.* p. 4.
16. N.K. Sinha, *The Economic History of Bengal, 1793-1848,* Vol. III, Calcutta 1970, pp. 14-16.
17. Gupta, *op. cit.*, p. 5.

HANDLOOM INDUSTRY

Closely allied to the production of silk and cotton was the weaving industry. In Assam every household had at least one loom and both spinning and weaving were considered a part of a woman's duties. The art of weaving had reached a considerable degree of perfection and very often cotton lengths were daintily woven with a lace of gold and silver thread. Sometimes cotton was also mixed with *Muga* or *Eri* to produce material for clothes. The main handloom products were *gamosa, bor kapor, churia, chador, khaniya kapor, cheleng, riha, mekela* and *paridiya kapor.* Although most of these articles were meant for home consumption, some of it was also bartered for other items with the neighbouring hill tribes. In the Brahmaputra Valley, Nagaon, Kamrup and Goalpara were the main districts where locally made cloths were sold in the weekly markets or in the occasional *melas* or *sobhas* that were held. A few important centres of the weaving industry were the Kadam Mauza of North Lakhimpur, Golaghat, Jorhat, Sualkuchi, Kamarkuchi of Hajo, Lashkarpara and Barpeta.[1] The major portion of the transaction was directly between the producer and the customer. However, at times middlemen, mainly Marwari traders, collected the cloths from the homes of the weavers and then hawked them about for sale. Trade in these materials was, however, brisker in the hills.

HANDLOOM INDUSTRY

Spinning and weaving in Assam were an integral part of the ethos of the people and everyone, irrespective of caste, creed or social status, was engaged in this industry. In fact it was considered an essential part of a girl's education. The weaving was done entirely in handlooms which were fairly simple contraptions. In an Assamese household, the loom was to a large extent the centre of the domestic economy, the only hope of salvation in an hour of distress or despair. If the rent was in arrear, and had to be paid immediately, it was to the product of the looms, worked by the women of the family, that the householder turned to.

Gradually, however, weaving as an industry began to stagnate. Initially, the mill finished products of England had met with certain obstacles in Assam which had delayed the efforts of the British to put wedges in the local handloom industry. But in the face of stiff competition from the British manufacturers the local industry started losing ground. In Assam, where cotton piece goods were the products of handlooms, the cheapest of labour could not compete either in quality or quantity with Lancashire mules or throttles. The local silk fabrics were replaced to a certain extent by the cheaper silk pieces from other regions. A length of Benarasi silk of 14×3 cubits, for instance, cost Rs. 32 whereas a similar length of *Pat* silk cost around Rs. 60.[2] In a society based on the market, cheaper products will naturally always prevail over the costly ones.

Earlier, the people had produced primarily for themselves and only the surplus was traded. Under the new conditions, the weaver became increasingly dependent on the *Mahajan,* the merchant, for the sale of his products in the market. Acute competition moreover, had necessitated an increased investment of capital which the local weaver rarely possessed. Hence, once again he had to fall back upon the *Mahajan* who more often than not provided the necessary capital at high rates of interest. This led to a situation where the weaver steadily came into his grasp. In such circumstances and in the face of an unequal struggle against modern machine-based industries, the Assamese handloom industry gradually declined. Significantly, however, despite all odds it survived....not as a flourishing industry but in a limited manner supplying the household necessities.

References

1. *A.S.R.,* Rev. Misc. B, 1909, April, No. 17-8.
2. *A.S.R.,* Rev. Misc. B, 1909, April, No. 61-2.

RHEA AND JUTE

Closely allied to silk and cotton were two other fibres, viz., *Rhea* and Jute. *Rhea* (Bohneria Nivea), which was grown locally in Assam. The *Rhea* fibre, also known as Ramie and akin to Hemp, is an extremely strong fibre suitable for manufacturing canvas, sail cloths, lines, ropes and cables. In Assam, the fishermen used the fibre for making fishing nets and ropes. *Rhea* was generally cultivated in small patches by the fishermen near the huts but some also grew wild at the foot of the hills.[1] Manure, moisture and shade were essential for the quick growth of the fibre to around 8 feet from which a fibre stock of 6 feet could be separated. The fibre was in great demand in England ever since the rapid growth of its shipping industry and most of it had to be imported from China, Russia and the Philippines. The Import Trade Returns of the United Kingdom showed a large and continually increasing delivery of hemp. In 1879, imports from the Philippine Islands alone had amounted to almost 20,000 tons valued at about half a million pound sterling.[2]

Dr. Forbes Royale, who was deputed to ascertain the properties and value of Assamese *Rhea*, commented that it was identical to Ramie and that "it was likely to prove one of the most valuable products of India, for in strength it exceeds the best hemp and in firmness it rivals the superior kinds of flax".[3] As early as 1803, Dr. Roxburgh of the

Botanical Gardens, Calcutta, had written that "it was one of the strongest fibres that he had met with" and that it could be easily propagated. Dr. Buchanan, also from the Botanical Gardens, sent three bales of the Assam fibre to M/s. Sharp Mark Lane, England, to be tested. The firm reported that the thread spun from this fibre bore 252 lbs. weight whereas the weight borne by the Russian hemp of the same size was only 84 lbs.[4] This view was corroborated by Marshall of Leeds, an authority on the fibre, who suggested the extension of the ordinary cultivation of the crop and the collection of the wild *Rhea* in sufficient quantities. He also urged that the fibre be prepared as carefully as possible for the English market. It was further pointed out that *Rhea* could be produced and sold with profit "at as cheap a rate as Russian hemp" and could speedily undersell all other fibres as four or five crops of the fibre could be obtained within the year from the same plant.[5] It was hoped that the *Rhea* from Assam would not only be sufficient to meet the British demand to a large extent but could be exported to other countries as well.

The main obstacle to the optimum utilisation of this fibre was, however, the difficulties faced in processing it. With a view towards developing the fibre of the *Rhea* plant as a regular industry, the Government of India offered in 1870-71 a prize of 5,000 pound sterling to the inventor of the best machine or process for its preparation.[6] Only one machine was presented for trial and this was found to be wanting in many respects. As the need for a good machine still existed, the Government renewed its offer of the prize in 1877. Seven machines were produced for trial this time but none of the fibres produced came up to the prescribed standard. In the absence of proper machinery, the government felt that *Rhea* plantations were not a viable proposition and was therefore not prepared to invest in such a venture until such time that "private enterprise has shown that the cultivation of the plant can be undertaken with profit".[7]

In the absence of outside interference, local production continued as before. The technology was archaic and the method of production an extremely slow and tedious process. Nevertheless the industry survived catering basically to the local demand.

The jute industry had a late start in Assam. Till about 1830, the manufacture of gunny bags and jute cloth had been the monopoly of the Bengal handloom weaver. Thereafter, an active manufacturing industry had sprung up in Dundee (Scotland) and it was found that it would be much more profitable to export the raw jute than produce gunnies on the handloom. The importance of jute as a raw material increased rapidly especially when, as a result of the Crimean War, supplies of Russian hemp were cut off. The increasing demand for jute necessitated the extension of cultivation and the government began to look beyond the borders of Bengal for probable areas. It was in this context that Assam came into the scene.

Prior to this, jute was grown by the *ryots* on a small scale on patches of homestead land or on the periphery of rice fields solely for domestic consumption in the plains districts of Assam and the plains portion of the Garo Hills. But as the interest of the government in the product grew, cultivation was encouraged on a commercial scale for the purpose of export, in the districts of Goalpara and Sylhet, adjoining the jute growing areas of Bengal. Jute thrived on clayey loam or sandy loam soil which was also suitable for the cultivation of *Ahu* rice. Basu, the Assistant Director of Agriculture, was of the opinion that as "jute is almost, if not twice as paying as *Ahu* rice", the Assamese *ryots* ought to be encouraged to cultivate the cash crop instead of rice in order to double their income.[8] F.J. Monahan, Director of Land Records and Agriculture, however, vehemently opposed this view because he felt that that sacrificing rice fields for jute cultivation would result in scarcity of food grains.[9] However, he felt that the Government should encourage the extension of cultivation by adopting such measures as granting waste lands, free distribution of quality seeds, enacting favourable immigration laws etc.[10] By the beginning of the 20th Century, production had considerably increased as jute cultivation had been taken up on a large scale by Muslim immigrants from East Bengal. In fact, the export of jute from the Brahmaputra Valley increased from 59,249 *maunds* in 1880-81 to 2,79,113, *maunds* in 1895-96.[11] In the Surma Valley, the increase for the corresponding period was from 9,787 *maunds* to 97,708 *maunds.*[12] The total quantity of raw jute exported from the province of Eastern Bengal and Assam ten years later stood at 24,256,000 *maunds.*[13]

It is intriguing to note that despite the fact that such large quantities of jute were produced in the region, no attempt whatsoever was made to set up a manufacturing industry in the province. Almost the entire quantity of raw jute was purchased by certain firms from the producers directly or through middlemen known as *Beparis.* These purchasing stations, which consisted of an office and godowns, were manned by the firm's own employees. The consignments were then shipped to Calcutta to be exported through their agents to feed the mills initially in Calcutta, and later, in England and Scotland.

The views of the Government on the question of setting up jute mills in the region were reflected by Dawson, an I.C.S. officer, in his Note on Jute.[14] He analysed that jute mills were not an economically viable proposition in Assam. He pointed out that since the main market for gunnies was in Calcutta, the mill owners in Assam would be handicapped from the beginning. He further stated that the absence of local labour and skilled technical personnel, created additional disadvantages. These are hardly acceptable arguments. When the tea industry in Assam could flourish despite greater difficulties, there was no reason why a manufacturing industry in jute could not have been successfully established. The mill labour in Bengal was almost entirely imported, and with sufficient inducements, it would not have been difficult to import labour into Assam. The tea industry in Assam had already proved this. Once the industry was established, skilled technical labour would also be available. The explanation, therefore, lay not in these reasons but was inherent in the colonial policy itself which discouraged any industry in India which was likely to provide competition to their own industrial products. Hence, with the bulk of the produce shipped out of the province, jute never developed into a full fledged industry in Assam at all.

References

1. Hence the name *Rihabari* given to specific localities in many places in Assam.

2. *Revenue and Agricultural Department, (Agriculture and Horticulture)*, July 1881, No. 29.
3. *A.S.R.*, File No. 283/612, 1854-57, Memorandum on the Production of Hemp and Rhea Fibre in India.
4. *Ibid.*
5. *Ibid.*
6. *Home Revenue and Agriculture Department, (Agriculture and Horticulture)*, May 1881, No. 22.
7. *Ibid.*
8. *A.S.R.*, Revenue A, July 1897, Nos. 45-47.
9. *Ibid.* A similar situation had occurred in China when the Chinese farmers started to cultivate opium in their rice fields.
10. *Ibid.*
11. *A.S.R.*, Report on River Trade for the Province of Assam 1881-82; Report on the River Trade of Assam for the year ending 31 March 1896.
12. *Ibid.*
13. *A.S.R.*, Report on the Trade Carried by Rail and River in the province of Eastern Bengal and Assam during 1907-08, p. 2.
14. Cited in Gupta, *op. cit.* p. 63.

LAC

Another agro-based industry of Assam was the cultivation of lac. The cultivation and collection of lac was almost entirely in the hands of the tribes dwelling in the poorer areas, and the methods of propagation and collection had remained unchanged for generations. These methods satisfied the local demand for the article which was used mainly as a dye. There was comparatively less demand for lac resin as its properties had not been fully ascertained and therefore, the export in lac was also mainly in the form of a dye. In fact at the beginning of the nineteenth century the value of lac dye was almost six times that of lac resin.[1]

Lac was reared in varying extent in almost all the districts of Assam, but the south bank of the Brahmaputra in the Kamrup district, Nagaon and the Garo Hills were the chief centres of cultivation. The lac was brought down by the producers to be traded at the weekly markets at Palasbari, Chaygaon and Boko. A small quantity was also brought down by the Bhutias to the annual fairs at Darranga. In Assam, lac was generally used by carpenters as a colour medium in the production of furniture, boxes etc., and by goldsmiths and silversmiths in their trades. The major portion of the produce, however, was exported. In 1883-84, a total quantity of 28,196 *maunds* was recorded to have been exported from the province.[2] In 1906-07, the quantity had increased to 45,028 *maunds*, valued at Rs. 22,49,629.[3]

Generally the lac was collected by people from the forest trees. The lac insect was also reared on several species of the *Ficus* family which were cultivated in *jhums*. The Garos usually raised their crop on the *Arhar* (*Cajanus Indicus*) and *Bolmango* (*Grewria Serabida*) trees. Generally the lac could be cropped twice a year but the second crop was raised primarily for its seed. The bulk of the produce was exported in the form of stick lac, i.e., the lac was collected by breaking off the twigs containing lac incrustation a few days before the larvae swarmed out. At this stage, the twigs are surrounded by deposits of a translucent orange yellow gum in which the insects are embedded. The gummy substance was then scraped off from the twigs and separated from the dead insects. The gum was then melted, strained, cleaned and sold as shellac or button lac.[4]

Initially, the producers themselves used to trade their lac in the markets or fairs, but gradually middlemen entered the field. These traders purchased the lac from the hill people to whom they lent money. More often than not, the simple villagers were ruthlessly exploited. Basu, Assistant Director, Agriculture, states that the *Mahajans* who made the loans often extracted up to three times the original loan amount.[5] All purchasers of lac were also required to take licences by paying a fee of Rs. 3 each and a tax of Rs. 2 to the Forest Department. The tax ultimately fell on the growers because the middlemen who bought the lac from them deducted this amount while making the payment.[6] Commenting on this point Basu wrote: "The duty, though levied on exports, falls ultimately on the growers. The *Keyas* at Lumding have been using the existence of the duty as a pretext for lowering the price of lac by 2 *annas* per seer, or Rs. 5 per *maund*".

The development of the textile industries in England had stimulated discoveries in industrial chemistry and with the introduction of synthetic dyes, lac dye was gradually robbed of its commercial value. But this was more than offset by the discovery of its new uses as a spirit varnish and polish, as a stiffening material for hats, as an ingredient for lithographic ink and as sealing wax, and demand for this product grew tremendously in Europe and America. Between 1868 and 1900 the value of the export of lac dye from India had increased from Rs.18,41,491 to 92,65,600 but the demand for lac stick increased tremendously.[7] The use of lac was likely to be extended rather than restricted and was not threatened by any rival. The substitute, which had been announced by Germany, had not yet been marketed on a large scale. Hence Britain began to view the cultivation of lac in Assam with renewed interest.

Unfortunately, the government left the initiative totally to private enterprise and did little to expedite the matter. Had the industry been taken up on a larger scale, the setting up of factories for the production of shellac would also have been feasible. Commenting on this, Stebbing wrote:[8]

> "It would seem a pity that the question of lac cultivation is not taken up more seriously in Assam. It would appear advisable for the Forest Department to take up the question with the object of showing what can be done by scientific management and by the introduction of better methods. The chief improvements required to be made are in the direction of the formation of regular areas of coppice either from seed or cuttings, which should be worked on a definite rotation.

Further, it is stated that lac reared on the *Cajunus Indicus*, which is said to be the best lac produced in Assam, can be put on other lac-rearing plants. Crops of lac could thus be raised in nurseries and distributed to inhabitants in the forests and villages. The formation of these lac nurseries is to be strongly recommended in Assam with a view to demonstrating their usefulness to the people, and thus improving the methods of collection and increasing the amounts available for export".

Stebbing's concluding statement that "it is almost inconceivably surprising that the cultivation of lac has been so neglected in the province" clearly sums up the apathetic attitude of the government.

References

1. *Maritime Trade Report for Bengal, 1907-08,* cited in Gupta, *op. cit.* p. 81.
2. *A.S.R.*, Report on the River Trade of the Province of Assam for 1883-84.
3. *Ibid.,* Report on the Trade Carried by Rail and River in the Province of Eastern Bengal and Assam
4. E.P. Stebbing, "Note on the Lac Insect', *Gazetteer of Bengal and N.E. India 1905,* p. 78.
5. Gupta, *op. cit.,* p. 82.
6. *A.S.R.,* Revenue, Misc B, April 1900, Nos. 17-18
7. *Ibid.*
8. Stebbing, *op. cit.*

GOLD WASHING AND JEWELLERY MAKING

Apart from the agro-based industries mentioned above, there were several other industries among which an important one was gold washing and the manufacture of jewellery. Nearly all the tributaries of the Brahmaputra were auriferous and gold was washed from the innumerable streams that intersected the land. Although the Dihing Bharali, Dikrang and Subansiri were the largest producers of gold, the smaller rivers like Tengapani, Parua, Diburu, Dekajuri etc. were also stated to contain a fairly good quantity of gold deposits.[1] Pemberton reported that "almost all the streams which flow into the Burhumpooter are in a greater or less degree auriferous, the gold obtained at the junction of the Burhumpooter and Dhunseeree river alone, was estimated by Buchanan, in 1809, at 1,80,000 rupees per annum".[2] The Muslim chronicler who came to Assam with Mir Jumla calculated that the annual value of gold washed from the sands of the Brahmaputra would be between Rs. 80,000 and Rs. 1,08,000.[3]

During the time of the Ahoms, gold washing was carried out by a guild known as the *Sonowal khel,* comprising of people usually drawn from the Kachari, Bihia, Koch and Keot communities, who paid the government a tax of five rupees worth of gold per annum. Individuals who washed gold, were required to pay one *tola* (11.66 gms.) per head per annum to the royal exchequer.[4] Captain Dalton, who visited the Jugloo and Sisi rivers with a team of labourers, found that by washing a ton of sand his men were able to procure about 18 grains of gold. According to him, fifteen days of hard work on the Sisi river would enable a person to procure enough gold to meet the revenue demand.[5]

Gold washing did not involve any sophisticated equipment. The process of gold washing has been described by Hunter thus:[6]

> "The process by which the gold is extracted appears to be similar to that used by diggers in Australia and California: the river sand is deposited in a rough wooden cradle by one man, a second operator keeps stirring the sand with his hands, while a third causes a continuous gentle stream of water to fall on the upper end of the cradle, by which all the lighter particles of sand are washed away, leaving only the heavier grains of gold".

The British initially leased out river *mahals* to the highest bidder for the collection of gold. At the beginning it brought in the desired returns, but soon it was found that the number of bidders was decreasing. Campbell attributed this to the *Omlahs* who apparently bought up the *mahals* at auctions and then made advance payments to the gold washers on the condition that the entire quantity of gold procured would be sold to them alone. The gold washers were thus trapped because they were tied down and were compelled to part with the gold at extremely uneconomic rates. At a time when gold was selling in the market at between Rs.16 and Rs. 17 per *tola,* the gold washers were forced to sell their products to the *mahaldars* at rates varying between 8 and Rs. 10 per *tola*.[7] In such circumstances, gold washing naturally became unremunerative and in course of time the industry gradually declined.

Although gold washing in Assam steadily declined, the goldsmith continued to occupy a prominent place in the village society. Apart from its status as a symbol of wealth, jewellery in Assam played an important role in ceremonies and rituals although crown or temple jewellery do not appear to be of any special significance. The tradition of crafting jewellery in Assam, is, therefore, a very old one. The Ahom monarch, Rudra Singha (1696-1714) is said to have imported goldsmiths from Benaras to train the Assamese jewellers.[8] The Census of 1881 returned 2,102 jewellers in the province of Assam while the Census of 1901 reported that 14,956 people were supported by this profession.[9] Although the local supply of gold gradually decreased, the demand appears to have been met by steadily increasing imports.[10]

The jeweller's work consisted of three different branches, viz., the preparation of ordinary gold and silver ornaments, the setting of precious and semi-precious stones in gold, and filigree work in gold and silver wire. The gold and silverwares were generally articles of personal adornment, though ocasionally items like small silver boxes (*tema*) or silver bowls (*bati*) were also manufactured.[11] Jorhat was the main centre for the enamelling of gold work. The enamel (*mina*) was usually of three kinds—a dark blue, a dark green and a white, but red and yellow were also used at times. Barpeta, in Kamrup district, was known for its artistic gold filigree work. The main items were silver and gold bracelets, rings, ear ornaments of various descriptions, beads and necklaces. Among the more popular jewellery items were the *Doog-Doogi, Loka-Paro, Bana, Gol Pota, Thuria, Gaam Kharu,* and *Jon Biri.* Henniker, in his monograph, comments that the jewellery of Jorhat were "neither very precious nor well cut", but that those of Barpeta were distinctly attractive and was sure to have a ready market if properly advertised.[12] H.Z. Darrah, while commenting on the gold work of Jorhat remarked that the craftsmen possessed "a fair amount of skill". But as they worked "almost entirely for the native trade, the articles produced lack the finish to be seen in ornaments manufactured for European customers".[13]

The first decade of the twentieth century saw an influx of jewellers from Dacca into the important towns of Assam where they set up a profitable business. The master artisan, who was usually also the owner of the shop, made a fairly decent income of between Rs. 50 and Rs. 80 per month, while his employees earned around Rs. 20 per month.[14] The artistic talents of these craftsmen had reached a high level of perfection "though they were inferior to the work of the Calcutta up-country shops like that of Tara Chand in Park Street".[15] In the face of competition, the local Assamese jeweller appears to have gradually given up his age old profession.

References

1. Fort William Revenue Progs, 17 Aug 1854, No. 15.
2. R.B. Pemberton, *Report on the Eastern Frontier of British India,* Guwahati, 1966, p. 84.

3. S.K. Bhuyan, *Anglo-Assamese Relations, 1771-1826,* Guwahati (reprint) 1974, p. 11.
4. E. Gait, *A History of Assam,* Calcutta 1963 (third edn) p. 272.
5. *Journal of the Asiatic Society of Bengal,* Vol. XXII, 1853.
6. W.W. Hunter, *A Statistical Account of Assam,* Vol. I, Guwahati (reprint) 1998, p. 380.
7. Kshetradhar Bargohain, 'Asamat Son' in *Awahan,* Vol. IV, 1933. cited in R. Saikia, *Social and Economic History of Assam (1853-1921),* New Delhi, 2000, p. 60.
8. F.C. Henniker, *The Gold and Silver Wares of Assam: A Monograph,* Shillong 1905.
9. *Census of India, 1901,* Vol.IV, Pt.1, Shillong 1902, p. 175.
10. In 1886-87, gold worth Rs. 9,187 was imported into Assam. In 1889-90, the amount had increased to Rs. 10,760. *A.S.R.,* Report on the Trade between Assam and the Adjoining Foreign Countries, 1890. Earlier Trade Reports do not include gold as item of import.
11. Henniker, *op. cit.*
12. *Ibid.*
13. Darrah and Gait, *Notes on Some Industries of Assam, 1884-1895,* Shillong 1896, p. 69.
14. Gupta, *op. cit.,* p. 56.
15. *Ibid.*

METAL WORK

Another notable industry was the metal industry. The workers in brass belonged to a particular community known as *Marias*. These *Marias* were supposed to have been the descendants of Muslim prisoners of war during Turbak's invasion of Assam around 1506 A.D. According to Assamese chronicles, they initially took to jungle cutting and then to cultivation, but having failed in both they took to metal work.[1] Highly industrious, the *Marias* were able to make a mark for themselves in their trade in which their womenfolk also took an active part. The *Marias* manufactured cooking and other household items of various shapes from well-amalgamated metal. They used sheet brass of about 1/16 inch thickness and joined various pieces together to produce the desired shape. The metal workers of Assam usually followed two methods of manufacturing their products. The more popular method, the hammering method, entailed beating out the plates into desired shapes with the help of hammers after heating them in furnaces. The component parts were soldered together by means of an alloy of copper, zinc and borax known as *pan*. The rough vessel was then turned in a lathe to be chiselled out. The second method entailed casting into specific moulds. Except the metal known as *bharan*, a mixture of copper and zinc, no other alloy appears to have been locally made in Assam. The braziers relied wholly on imported brass plates to make their utensils. In 1907–08, a quantity of 66,000 *maunds* of brass and copper valued at Rs. 30.9 lakhs, was imported into the province.[2] The *Marias* normally surveyed an area before deciding to settle down in a particular village.

Bell metal items were usually made by the *kahars* who were "mostly persons of *Kalita, Koch* and *Kurat* castes and the word *kahar* being tagged on to their proper caste name as an indication of the occupation they follow.[3] The *Kahars* generally manufactured their articles by melting down old items.[4] Gait noted that there were around 400 families of *Kahars* in Sarthebari, the main centre of the bell metal industry, of whom around 100 families were dependent solely on this occupation. The rest took up the profession as a subsidiary occupation.[5] Some *kahars* occasionally migrated to neighbouring Guwahati and Barpeta temporarily to carry on their business. Common bell metal articles were *kahi* (plate), *bati* (bowl), *lota* (jug), *kalah* (pitcher), *gagori* (pitcher), *sarai* (tray) etc.

Copper was not widely used. Its use was limited to the issue of copper-late inscriptions by kings and decorations in temples.

The main centres of the metal industry were Sarthebari, Hajo, Baligaon,Titabor and Guwahati. The majority of the metal workers, however, suffered from the handicap of not being directly in touch with the consumers. The industry was almost entirely in the hands of the *mahajans* who supplied the workers with the raw materials and took back the finished products from them after paying them wages at fixed rates per *maund* of manufactured article. In some cases there was a second class of middlemen, known as *beparis,* who brought the metal and money from the *mahajan* and advanced them to the workers. In either case, the profits went largely to the middlemen.[6]. An artisan could turn out between one and one and a half *maunds* of utensils per month. Although the price of the raw material was around R. 60 per *maund* and the wholesale price of the manufactured articles between Rs.150 and Rs. 170 per *maund,* the artisan ended up earning only around Rs. 30 per month.[7] The majority of the metal workers did not have the requisite capital at hand to set up their own business. As such, they had no alternative but to depend on the *mahajans* for their livelihood.

In course of time, the local products began to face stiff competition from imported enamel ware and china vessels. As most of locally made items had to be manufactured from imported material, the price was considerably higher than the imported manufactured varieties. Moreover, the demand for enamel ware or china ware increased steadily among the Muslims and the immigrant tea garden labour population of Assam. Consequently, local production gradually declined. Despite great odds, however, the industry did not die out completely. The majority of the Assamese were Hindus and they were averse to the idea of substituting their age old traditional bell metal and brass utensils with the imported items. The industry, therefore, survived for some time, though to a limited extent only.

Another branch of the metal industry was that of the blacksmith. The blacksmiths or *kamars,* who generally belonged to the *Koch* or *Kalita* communities, manufactured household articles like knives, sickles, spades etc., and weapons of war like *hilloi* (matchlocks), *top* and *bor top* (small and big cannons), swords, spears, short javelins, and tips of arrows, among other items.[8] Shihabuddin Talish wrote that the Ahoms "cast excellent matchlocks and *bachadar* artillery and show great skills in this craft".[9] Along with firearms, the manufacture of gun powder was an equally important industry during the time of the Ahoms. In 1662 Mir Jumla is said to have found more than 4,260 *maunds* of gunpowder at Garhgaon and was so impressed with the quality that he is said to have ordered the capture of some expert gunpowder manufacturers.[10] The use of bows and arrows in war resulted in considerable growth of this industry. The bows were made of bamboo and the arrows, also of bamboo, were fitted with an iron head. The bow makers were organised into a guild known as the *Dhanu chausa khel* and were required to supply a certain number of bows annually to the royal store.

Iron ore was found in several places in Assam. In the Khasi Hills, the ores were in the form of fine sands and were irregularly distributed in the mountains. After collection, these were washed out of the sand, melted in a primitive furnace and then converted into balls from which weapons of war or agricultural implements were manufactured.[11] The quality of the iron of the Khasi Hills was reported to be 'excellent"[12]. In Uppet Assam, iron was found in nodules of various sizes and sometimes in thin layers in the coal-beds of the Naga Hills. A certain quantity was also found near Dergaon,[13] but no iron was smelted in Assam.

The volume of the industry of the local blacksmith is reflected in the import of raw iron into the province. The trade figures for 1895-96 give a total of 1,87,977 *maunds* of iron valued at Rs. 10,10,376 being imported into the Brahmaputra Valley alone.[14] The total quantity imported into the province during the same period was 2,85,527 *maunds* valued at Rs. 15,34,707.[15] There was very little competition with imported articles for the common agricultural implements and household articles that were used by the villagers. The price of the raw material was also reasonable. Iron could be bought at around Rs. 5 per *maund* and steel at Rs. 13 per *maund.* The finished product was sold from Rs.1-4-0 to 2-8-0 per *seer.*[16] In most cases, steel was used sparingly, only a minimal quantity being welded to the edge of the implements in order to harden and sharpen them. The productions costs were thus kept low. It, therefore, appears that the profit in this industry was greater than that of any other village industry. The ordinary blacksmith earned around Rs 15 per month, and since the industry was free from the interference of middlemen, those engaged in the industry had little to complain about.[17]

British rule inaugurated an era of modernisation and was characterised by technological advances and innovations in the industrial and communications sectors. Significantly, in spite of technological innovations in the tea, coal, oil and other industries and in the construction of railways, the work of the local blacksmith remained confined to manufacturing the traditional household implements. Machinery for tea factories, rails and other gear for railways, enamelled ironware for the agriculturist and galvanized iron sheets for buildings, were all imported from outside the province. As a result, the low technical skill of the village blacksmith had little scope for improvement.

References

1. H.K. Barpujari ed, *The Comprehensive History of Assam*, Vol. V, Guwahati 1993, p. 104.
2. *A,S,R.*, Report on the Trade Carried by Rail and River in the Province of Eastern Bengal and Assam, 1907-08, p. 2.
3. Darrah and Gait, *op. cit.*, pp. 112-16.
4. Gupta, *op. cit.*, p. 20.
5. Darrah and Gait, *op. cit.*, pp. 112-16.
6. *Ibid.*
7. *A.S.R.*, Rev.B, March 1903, Nos. 845-65.
8. Francis Hamilton, *An Account of Assam 1807-14*, London 1820, (reprint , Guwahati 1940), p. 60.
9. *Fathiya-i-Ibriya*, J.N.Sarkar's translation cited in J.N. Phukan, '*Economic History of Assam under the Ahoms*', Gauhati University, 1973. (unpublished thesis) p. 122.
10. J.N.Sarkar, *Life of MirJumla*, Calcutta 1916, p. 249.
11. Barpujari *op. cit.*, p. 77.
12. Gupta, *op. cit.*, p. 45.
13. *Foreign Political Proceedings*, 11 Feb. 1835, No. 90.
14. *A.S.R.*, Report on the River Bourne Trade of Assam, (Year ending 31 March, 1896), p. 7.
15. *Ibid.*
16. Gupta, *op. cit.*, p. 43.
17. *Ibid.*

TIMBER

Assam possessed extensive forests. The diversified elevation of the land yielded a variety of trees some of which were very valuable. The *Som* (*Artocarpus Chaplasha*) and Savannah forests were found in the plains, and the *Sal* (*Shore a robusta*) forests partly in the plains and partly in the lower hill forests to an elevation of about 3,000 feet. The pine forests were from 2,500 ft. to 6,000 ft in the upper hill areas.[1] The *Sal* forests were of considerable value owing to their timber, while the *Som* forests were important on account of sericulture. The mixed and lower hill forests were rich in cane, bamboo and a variety of deciduous trees including caoutchouc (India rubber). The nearness of the forests to streams and rivers provided easy access to the Brahmaputra on which the main trading stations were situated. Hence, an active industry in timber existed in Assam. The *Sal* and *Som* were used mainly for boats and planks, *Ahjar* (*Logerstroemia regina*) for boats and buildings, *Nahor* (*Mesua ferrea*) for posts, *poma* (*Cedrela toona*) for furniture, planks and tea boxes, *Titasappa* (*Michelia campaoa*) for boats and furniture, and the rest to meet the general requirements.[2]

(i) Boat Building

The physical features of the land largely contributed to the development of the boat building industry in Assam.

Since the Brahmaputra Valley was intersected by innumerable streams, boats formed the most convenient and fast form of communication and transport in the province. Moreover, the forests were full of hard and durable timber suitable for boats and thus a considerable industry existed in canoes hollowed out from large trees. The boats were of varied shapes and sizes. M'Cosh commented that in Assam "a boat (was) as common to every household as a brass *lota* or an earthen pot".[3] The common man built his own boat, whereas the State employed several thousand *Naosalias* for the purpose.

The Brahmaputra was the only highway which connected the province with the rest of India till the introduction of railways. The Commission of Enquiry appointed by Lord Bentinck to visit Assam in 1834 had taken four and a half months to reach Sadiya from Calcutta.[4] M'Cosh, writing in 1860, stated that a large boat took between six and eight weeks to reach Guwahati from Calcutta though the post, conveyed in small canoes, reached Guwahati in ten days.[5] As the only route of travel to and from the province was by river, boat building was an important industry of Assam. In fact it remained so for a long time even after steamer services on the Brahmaputra and railways in Assam were introduced. Steamer and railway charges for freight were initially fairly high. When the Indian General Steamer Navigation Company started operations on the Brahmaputra in 1860, a ticket from Calcutta to Guwahati cost Rs. 150 whereas freight on ordinary stores was at the rate of Rs.1 per cubic foot.[6] Moreover, as the services were not very regular, the steamers could never be relied upon for the despatch of goods. Hence, for a long time, the greater part of the trade of the province continued to be carried by country boat. Most of the tea companies were compelled to maintain their own fleet of country boats and, having got the boats, to use them. Building up a private fleet was, of course, no mean task. There are references to small native boats of 300 *maunds* capacity costing Rs,. 250 each.[7] Hire charges for a boat of 300 *maunds* capacity along with the crew, was Rs. 37 per month.[8] Despite this, in the absence of a developed communication network, river transport continued to be of great importance for a long time. Pandu, Guwahati, Kaliabor, Bishwanath, Dikhowmukh, Rangpur, Garhgaon and Sadiya were the main centres of the boat building industry.

(ii) Timber Posts and Furniture

The carpenter was an important person in the village and in considerable demand. Generally also an agriculturist, he catered to the needs of his fellow villagers. He normally possessed some basic tools like a saw, a chisel and crafted implements such as ploughs, *dhekis* for husking rice, *ghanis* for the oil mill, looms for the weavers, cart wheels, doors and windows and some furniture. Among the carpenters there were some who were skilled in the art of wood carving. This skill was confined to certain families only and the art was carried down from father to son. Most places of importance like public halls and *Naamghars* had decorated wooden posts, panels, doors and windows. These craftsmen, known as *Khanikars,* were also adept at making boxes, trays, stools, chairs and similar items, often carving them out from single blocks of wood.

Wood carving was normally seen in places of worship like *Naamghars* and *Kirtanghars.* The *Satras* (Vaisnavite monasteries) were important centres of this art where the *khanikars* depicted not only religious themes but also popular scenes, like bullfights, in their carvings.[9] However, the focal point of a village *Naamghar* or a *Satra* was the intricately carved *simhasan,* the multi-tiered raised platform on carved lions. They varied in size and quality but were all made from good durable timber. The *simhasans* of the Dakhinpat and Auniati *Satras* were inlaid with gold and silver. Gait remarked that "objects used for religious purposes will, in all probability, continue to be made as long as the Hindu religion retains its hold".[10]

The carpenters usually made goods to order, the price varying according to the nature of labour involved. The price of a *simhasan,* for instance varied between Rs. 25 and Rs. 150, a bed cost from Rs. 20 to Rs. 40 while a *dola* (palanquin) could be had for around Rs. 30.[11] Other products of the carpenters for general use included the *gasa* (lamp stands), *jatar* (spinning wheels), *kharam* (*wooden* sandals), *Borpira* (stool), *Salpira* (bed) etc. A few skilled wood carvers worked on wages varying from Rs. 15 to Rs. 30, but most carpenters took up the trade as a subsidiary occupation. As there was no scope of an export trade in their products, the market was not regular or consistent enough to ensure financial support to the artisans. Hence, they were compelled to rely heavily on their primary occupation, viz., agriculture.

As Assam was opened up to the rest of India there was an influx of immigrants into the province. A sizeable number came from Punjab, not as industrialists but as traders and carpenters. The local carpenters, with their low level of expertise, found it difficult to compete with the outsiders. Gradually, these skilled craftsmen superseded the local artisans in their trade to set up profitable business in woodwork in Assam.

References

1. *A.S.R.,* File No. 38/45, 1873, No. 94C; W. Schlich, *Memorandum on Forest Operations in Assam.*
2. *A.S.R.,* Forests, File No. 57/95 of 1873.
3. J.N.M' Cosh, *Topography of Assam,* Calcutta, 1837, p. 28.
4. Antrobus, *op. cit.,* p. 66.
5. M'Cosh, *op. cit.,* p. 82.
6. B.C. Allen, *op. cit.,* p. 225.
7. Antrobus, *op. cit.,* p. 349
8. *Ibid.*
9. Naren Kalita, *Bordowar Shilpabastu,* Nagoan 1985, p. 14.
10. *A.S.R.,* Rev, A, 1894, May, Nos. 91-97.
11. Barpujari, *op. cit.,* p. 103.

POTTERY AND EARTHENWARE

The potter was a common figure in every village. Almost all the potters were primarily agriculturists and took to the trade as a secondary occupation only. The *Kumars* and the *Hiras* were the two potter communities in Assam.[1] While the former used the potter's wheel, the latter did not use the wheel at all. But in general, the technology used by the Assamese potters was very crude. In his Report Gupta writes:[2]

> "The wheel used is the old solid wheel, which does not revolve as fast as the Bengal wheel, with a heavy periphery attached by spokes to the central hub. The firing and moulding are also very poor. The earth used is glutinous clay."

Apart from regular household items, like pitchers, plates etc., rings for wells, drums, pipes and toilets were also manufactured by the potters in considerable quantities. The quality, however, was rather poor. Significantly, the local potters did not appear keen to make any improvement in the quality of their products. They reasoned that since the religious observances of the Hindus require them to occasionally destroy their pots, the thin flimsy pottery of the province (priced 1/2 *pice* to 3 *pice* per ware) served the purpose and that in the circumstances there would be no demand for any superior quality expensive pottery.[3] This was hardly a valid argument. The demand for durable pottery among the tea garden labourers was considerable and they were even prepared to pay a high price of between 3 and 8 *annas* per ware for quality products. In fact boatloads of pots, mainly cooking pots and pitchers made their way frequently from Bengal to Dibrugarh.[4] There were also many instances of potters from Bihar having set up profitable business outside the tea gardens catering to the needs not only of the tea garden labour population but also of the neighbouring villages. The Assamese potter, apparently unmindful of the economic shift around him, let the opportunity slip by. He had no one to blame but himself for his economic backwardness.[5] His counterpart in the Surma valley, who also made clay images of various idols apart from the regular household items, was slightly better off. But even there a large amount of pottery was imported from East Bengal.

It is thus apparent that in Assam the industry in pottery was a stagnating one with little scope for improvement. Changed economic conditions, moreover, led to the gradual replacement of earthen ware by the imported enamel ware and chinaware. Although these were initially more popular among the immigrant population, they were able to break through the barriers in course of time and penetrate into virtually every household.

References

1. Gait was of the opinion that the *Hiras* in all probability were *Chandals* while the term *Kumar* did not denote any

particular caste but was applied to *Kalitas, Koches* and *Keots* who themselves or their ancestors were potters.

2. Gupta, *op. cit.,* p. 54.
3. *Ibid.*
4. *Ibid.,* One standard boatload of pottery usually fetched around Rs. 1,000.
5. In Assam, a potter earned around Rs. 4 per month as compared to Rs. 10 in Dacca. *Ibid.*

BONES, TUSKS, HORNS, HIDES AND FURS

Assam being rich in fauna, an impressive industry existed in bones, tusks, hoofs, hides and furs. The most prominent among these was undoubtedly ivory and rhino horns and hide. Rhino hides were primarily used as war armour. Hides of other animals were generally used to make mats. It is of significance that the art of treating and preserving the hides had reached a considerable degree of proficiency. Almost 85 per cent of the hides exported from India passed through Calcutta and Assam contributed a substantial portion of this trade. The trade figures for the years 1891-92 and 1892-93 show the quantity of exported raw hides from the Brahmaputra Valley as Rs.4,810 and 7,732 *maunds* respectively.[1] The value of exports in hides increased substantially from 2,17,000 in 1890-91 to Rs 5,65,000 in 1900-01 and Rs.9,32,000 in 1903-04.[2]

The method in which the trade was carried on has been described by Watson.[3] *Chamars* sold the hides to Muslim traders who exported them to Calcutta. Besides the local collecting agencies, a considerable quantity of hide was also collected by European exporting houses through their own agencies in the districts. The price paid to the *chamar* for 20 lbs. of medium quality hide was Rs. 11-4 *annas* whereas the same fetched Rs. 16 at Calcutta. The collector's profit was thus over 33 per cent. In pre-colonial times and during the early years of colonial rule, the *chamar* got the carcasses of dead animals free of charge from his fellow villagers. But once the demand for hides rose, the owners of the dead animals found it profitable to sell them directly to the representatives of those involved in the trade which meant that the *chamar* was deprived of a profitable source of income.

The quantity of leather tanned locally was very small and was used for the "roughest kinds of articles".[4] Practically the entire quantity of hide that was available in the province was exported in a raw state. Although most large villages had their own shoemakers, the major portion of the leather goods consumed in the province was imported. In 1899 the quantity of raw hides exported from Assam was over 40,000 *maunds* while the quantity of manufactured leather goods imported was around 3,000 *maunds*. There are no separate figures to show the quantity of footwear imported into Assam as the figures are included under the general head "Apparel".

The general trend of industrial development in Assam was also reflected in the leather industry. It is, therefore, not surprising that both in volume and trade, the export of hides and skins was by far the most important part of the industry. There were no proper tanneries in Assam and, therefore, the quantity of leather tanned locally was very small. Had the leather been tanned and manufactured locally, instead of being exported in a raw state, the product would have been of much greater value.

The same was the case with ivory. Some ivory was traded with Bengal in the form of manufactured goods such as boxes, handles, combs, jewellery, stools, etc., but the major portion was sold as tusks by Marwari traders who acted as middlemen between the actual hunters and the exporters in Calcutta. Trade figures for the year 1892-1893 shows 51 *maunds* of ivory being exported from the Brahmaputra Valley.[5]

References

1. *Ibid.*
2. Allen, *op. cit.*, p. 127.
3. Watson, *'Report on the Leather Industry in Bengal'* 1907, cited in Gupta, *op. cit.*, p. 47.
4. *Ibid.*
5. *A.S.R.*, *Report on the River Trade of the Province of Assam*, 1992-93, p. 17.

BAMBOO AND CANE

Bamboo and cane formed an integral part of the economy and lifestyle of the Assamese. Both are so versatile that they were the two most widely used products in Assam. It is, therefore, not surprising that a flourishing industry existed in these articles. They were used for a variety of purposes that included, among other items, the manufacture of household implements, construction of dwelling houses, production of weaving accessories and the manufacture of furniture, toys, ornaments and musical instruments. Bamboo was also very effectively used as a medicine. No community was exclusively connected with this craft and, like the weaving industry, the production of cane and bamboo products was generally carried out by all as a household industry. Although some substitutes for bamboo and cane gradually appeared in the market, the versatility of the two was such that the industry was able to survive in spite of competition from different quarters.

MASKS

With the tribal art and folk elements forming a dominant factor in Assamese culture, masks found an important place in the cultural activities of the people. Made from a variety of articles such as terracotta, pith, metal, bamboo and wood, masks were extensively used in *Bhaonas* and *Jatras* (folk theatre). The use of masks was varied and widespread but the thematic content of masked performances was generally drawn from the vast corpus of legends and revolved round mythological characters. The masks, with their manifold form of guises, was a very significant form of cultural expression. It was regarded as an instrument of revelations, giving form to the formless and building a bridge between the mundane world of everyday experience and the supernatural world of the gods and sacred entities. Apart from use in cultural performances, masks have a ritualistic significance for certain communities. The Tiwa community of Assam, for example, use masks of birds, animals and Lord Siva while performing the rites of the *Barat Puja.*

The *Ankia Nat,* the theatre form created by Sankardeva (1449 A.D.-1568 A.D.) to spread the message of *Vaisnavism,* were one-act plays characterised by glittering costumes and fanciful effigies of demons and monsters. These huge effigies, made out of bamboo and covered with paint and *papier mache,* were sometimes as tall as 15 feet and needed several actors to manipulate them. Animal masks were also worn. The actors in these masked performances danced to the rhythm of music which obviously meant that the masks had to be light weight unlike masks in many other dance forms. Generally there were three types of masks, viz., *mukh mukha, cho mukha* and *latokai mukha.*[1] The first was worn over a face. The *cho mukha,* a mask which covered the entire body, had to be kept in position with the helps of ties. The *latokai mkha* was similar to the *cho mukha* but smaller in size.

The *bhaona* masks were made of fine bamboo strips and then covered with a paste of cow dung mixed with soil. After being dried in the sun they were painted with bright indigenously prepared colours. In course of time, as regional variations began to develop within the *bhaona,* the size and nature of the masks also underwent a change. The *Khuliya-bhaoriya* of Darrang district, for instance, while retaining the essence of the *bhaona,* specialised in small wooden masks while discarding the bigger ones. In Goalpara and Dhubri districts the rich traditions of folk paintings on *solapith* has also found reflection in a variety of painted masks.

Mask-making in Assam has always been a male dominated profession. It is a skilled art which was generally passed down from generation to generation in particular families. The skill, however was usually confined to only those involved in the *satra* organisation and few outside the organisation were known to have any knowledge of the intricacies of the art. Mask-making, however, was just a subsidiary occupation of those engaged in its production. Like most other craftsmen in the province, they, too, were primarily agriculturists.

In course of time as the importance of *bhaonas* and *jatras* declined, mask-making also lost its importance. Today, mask-making is no longer the vibrant industry that it once used to be and the few masks that are manufactured are normally projected as showpieces of Assam's past cultural heritage.

References

1. Medhi and Zaman, 'Mask Makers of a Vaishnavite Monastry of Assam', in *Journal of Northeast India Council of Social Science Research*, Vol. 28, No.1, April, 2004.

DOMESTIC ECONOMY - A VIEW

In the pre-colonial period, the artisan industries constituted the industrial component of a balanced self-sustaining economy. In a predominantly agricultural set up where the main occupation of the people was cultivation, the people took to the various crafts as subsidiary occupations only. Hence there was no marked division between the capitalist and the labour. Within the artisan industry itself, there was not much specialisation as most workers practised their craft only in their spare time. Moreover, in Assam, as no worker was hereditarily attached to his trade, his level of technical skill was rather low and in the absence of external competition, the scope for improvement was minimal.

The Assamese were simple people with limited wants. The domestic economy had been based largely on barter, and money had played a significantly minor role as a medium of exchange. Moreover, the *paik* system, which had involved payment in the form of labour, did not foster a money economy. The savings of the well-to-do people were generally in the nature of gold and silver ornaments. In the days immediately preceding British occupation of Assam, the province had gone through a dark period of revolts and foreign invasions involving wanton destruction and looting. Much of what had been saved in gold was thus lost in the process. In any case, in all self-sustaining non-monetised village economies, where both production and market are severely restricted, the scope for a surplus is limited. In the Assamese society where the economy was based on barter, only those forms of production were selected which were necessary to satisfy the basic human wants with a minimal amount of labour. Moreover, in Assam, an additional factor was in operation. The general isolation of the region, owing partly to her geography and partly to the isolationist policies of the Ahom rulers, had resulted in an insular attitude among the Assamese.

The situation, however, completely changed within a few years of British occupation. British rule initiated a process of completely restructuring Assam's economy. The earlier *paik* system was initially replaced by a poll tax of Rs. 3 per *paik.* This was soon abandoned in favour of a regular assessment on land revenue. By 1840, the mode of collection had been fully monetised, the tax being payable by individual landholders. But assessments were made without an accurate survey of the land and, as a result, the tax levied had little relation to the area under cultivation. The introduction of a monetary economy placed the Assamese cultivators under severe economic strain. The land revenue of the Brahmaputra Valley, which amounted to Rs. 6,14,322 in 1842, rose to Rs. 7,43,689 in 1852 and jumped to Rs.21,65,157 in 1872-73 and to Rs. 41,27,328 in 1890-91.[1] This naturally fell heavily on the peasants. With the limited paying capacity of the cultivators, the land revenue was transformed into a virtual rent-rack which impoverished them.

As Britain systematically converted Assam into a raw material producing hinterland, the process of driving out the people from their crafts also proceeded at an alarmingly fast rate. This meant that the people were deprived of their subsidiary occupations and were compelled to rely heavily on their lands. Gait is of the opinion that the main reason for the swift decline of the traditional industries in Assam was the fact that in the absence of an occupation based caste structure, the artisan lacked the support system which is so essential for the survival of an industry. While agreeing that this was perhaps one reason, it is important to remember that, as A.R. Desai has remarked, "the historical strength of the machine industries lay in the fact that its products were cheaper than those of the handicrafts. In a society based on the market, the cheaper products will always prevail and oust the costly ones".[2] Many of the indigenous industries of Assam died under the pressure of new forces and those that survived remained undeveloped and primitive in organisation. The worst effect of this increasing loss of traditional industries and the failure of new industries to take their place was that the economy of the province came more and more under foreign domination.

The unequal competition stifled most of the local industries. An absolute dearth of accumulated capital or current savings among the people at large resulted in a situation wherein they were not even in the position to buy up the different *mahals* that were put up for auction, let alone set up individual industrial enterprises. In Assam there were no local merchants or traders who had amassed wealth in commercial pursuits. For the majority, agriculture was the main occupation, while the educated section preferred white collar jobs. So far as business was concerned, the *Keya mahajan,* like an octopus, was "omnipresent and filled up all profitable avenues in trade and commerce throughout Assam". They were, however, neither industrialists or entrepreneurs, but only traders. In the circumstances, foreign capitalists monopolised and reaped all the advantages of Assam's material resources. A notion widely prevalent in England and encouraged by the official circle in India was that India was never destined to be a great industrial country. Her role as a tropical colony lay in producing raw materials only in order to be utilised by Western countries possessing superior technical and scientific skill[3]. British commercial policy in Assam was also guided by this general attitude. As in the rest of India, here also the colonial government encouraged only those industries which were unlikely to rival their own industrial products.

Years of uninterrupted political and economic forces that had been in operation in Assam resulted in a situation where the indigenous crafts and industries of Assam had become a relic of the past order. The artistic excellence and inherent beauty of some of the products had never been disputed[4] but the question arose whether, in utility and cheapness, the revival of the indigenous industries would be economically viable. In certain quarters, it was strongly felt that the government should assist and develop certain indigenous industries which were likely to have a future and which were commercially possible to revive. There was, however, a note of caution. "There should be no avoidable false starts and no energy should be wasted in wrong and impossible directions. The raising of this point is not a mere platitude, for it is most important that whatever Government is able to render, must be wisely given"[5].

It was believed that if certain industries could be revived, the exporters of these articles could "attempt to secure markets in other parts of Asia, like China, instead of attempting to compete with European manufacturers in their own homes".[6] The idea behind this proposal was also to attract Indian capitalists with whose collaboration the English merchants could set up a profitable trade in these products. Regeneration of old industries was, however, just one aspect of the new policy. The other was an attempt to mobilise new resources which "would end in our securing a diversion ....of industrial enterprise and the establishment of those new methods of industrialism which now command the commerce of the world".[7]

Hence, towards the close of the nineteenth century, the attention of the Government was drawn towards regenerating industries like silk and sericulture, cotton, gold–washing, leather and metal work. The mobilisation of new resources included the commercial exploitation of rubber, *rhea*, jute and lac. The regeneration of industries proved a difficult task. As the official report stated:[8]

> "The present decadent condition of indigenous industries is the result of many economic and political causes, which have worked uninterruptedly for many generations and it will be no easy task to arrest the disruption which has followed. For, while Assam has lost her own industries, other nations are in possession of her markets, and the great start which they have gained, coupled with their manifold resources in modern methods, machinery and almost unlimited capital will make the struggle a most unequal and difficult one".

It was universally recognised that the cheapness of machine made products had largely contributed to the decadence of the artisan industries. If a fresh lease of life was to be given, the Government would have to be prepared to devise ways and means by which the local methods of production could be improved and made less expensive. It was, therefore, necessary to ensure that only those industries were encouraged whose products had an adequate local demand and for which raw material and labour was available in plenty. Another important factor was the question of local capital investment. The dearth of accumulated capital among the Assamese was apparent and industrial development of the province had so far revealed an absence of Assamese industrial enterprise. The Government identified the following factors for the prevailing situation:

(i) Insufficiency of fluid capital;
(ii) Absence of skilled supervision in technical matters;
(iii) Want of trained labour;
(iv) Lack of knowlwdge of business methods.[9] It was hoped that Government assistance on these matters would help to rectify the situation.

In the following years, the Government did make some attempts in this direction, but in most cases the attempts were half-hearted. Committees were formed to survey the situation and put forth their suggestions. Beyond this, the Government took little positive action. This attitude was reflected very clearly while the Government was toying with the idea of regenerating the leather industry. Cumming, in his Note, had summarised the following requisites for the successful operation of the tanning factory: (i) a hide market; (ii) supply of raw material for tanning; (iii) good water supply; (iv) labour; and (v) a local colony of *chamars*. Most of the above mentioned requirements were available in Assam, Even excellent quality tannin was locally available from certain trees like the *Sal* tree. Cumming also pointed out that the vast influx of immigrants into the Brahmaputra Valley had led to a tremendous increase in population. By 1902–03, the total imported labour force alone accounted for 13 per cent of the total population of 26,19,100 in the Brahmaputra Valley.[10] This itself assured a ready market. In spite of such a positive report, no initiative was taken to set up a factory in the region. It is thus apparent that the Government's intention was to allow the commercialisation of products only in those sectors and to the extent that did not adversely affect, in any way, the British industrial interest.

One recurring theme in all the Government reports was that non-availability of local labour proved the most serious impediment towards the successful operation of any project. This was undoubtedly true to a certain extent. However, it needs to be pointed out that even local labour could have been attracted with proper wages. In Lower Assam, for instance, Alymer, the Conservator of Forests, had succeeded in making extensive plantations of teak, rubber, *toon* and *sissu* entirely with local labour. By paying an attractive rate to every labourer who came and by giving the men perfect liberty to leave when they desired, he was able to obtain as much labour as he wanted.[11] This apart, it must be remembered that if the entire tea industry of Assam could be worked with imported labour, there was no reason why the same could not be the case for other sectors.

The retarded economic growth of Assam was, however, not an isolated phenomenon. It formed part of the all India pattern under British dominion. But the additional handicap that Assam faced was that of being geographically isolated from the rest of India and the lack of an effective communication network to link her to the mainland. Yet, despite these handicaps, flourishing tea, coal and oil industries had been set up. But owned, managed and worked by outsiders, as these sectors were, they had marginal impact on the local economy. Like all developing countries, here also the problem lay not in the creation of wealth but in the capacity to create wealth.

It was under the shadow of these drawbacks that Assam emerged out of colonial rule in 1947. With independence, efforts have been made to put back Assam on her tracks. The beginning of the first Five Year Plan initiated the process of economic development in Assam. But even today, an economic disparity in Central aid exists between Assam and the rest of India and this has led to much resentment. With government sympathy and aid, Assam, with her tremendous economic potential, can indeed be one of the most economically advanced regions in the country. The population of Assam is primarily rural. Therefore, what is required is an industrial resurgence in rural Assam if the state is to prosper. Small craft-type enterprises in the informal sector and small and medium enterprises in the organised sector can play an important role in uplifting the economy of the state.

References

1. *A.S.R.*, (General Department) Revenue A, April 1902, No. 16-75-2; A Guha, *Planter Raj to Swaraj: Freedom*

Struggle and Electoral Politics in Assam 1826-1947, Delhi 1977, p. 10.

2. A.R. Desai, *Social Background of Indian Nationalism, Bombay,* (reprint 1994), p. 97.
3. Bipan Chandra, *The Rise and Ggrowth of Economic Nationalism in India*, New Delhi, reprint 1982, p. 72.
4. W.W. Hunter, *Imperial Gazetteer of India*, Vol. IV, p. 572.
5. *A.S.R.*, (General Department) Revenue A, April, 1902, No. 43.
6. Gupta., *op. cit.*, p. 102.
7. *A.S.R.*, (General Department) Revenue A, April 1902, No. 43.
8. Gupta. *op. cit.*, p. 108.
9. *Ibid.*
10. B.C. Allen, *Imperial Gazetteer of India*, 1908, p. 125.
11. *A.S.R.*, *Report on the Forest Resources of Assam*, 1908, p. 20.

SECTION II

PROSPECTS OF INDIGENOUS INDUSTRIES

SERICULTURE

Assam is ideally suited for sericulture as the region possesses all the natural conditions conducive for rearing a variety of silkworms on a commercial scale. We have mentioned earlier that, since time immemorial, a flourishing industry has existed in the manufacture of *Pat, Muga* and *Eri* and that these have been a part of the cultural heritage of Assam. At present, Assam is practically the sole producer of *Muga* silk in the world (99 percent). Efforts, however, are being made to produce the silk on a commercial scale in parts of China as well as in a few other states of India. The contribution of Assam in the production of *Eri* in India is around 54 per cent, while that of *Pat* is negligible. It is ironical that this is so despite the fact that the production of *Pat* has increased almost 23 times since 1950 whereas production of *Eri* has increased only 7 times and has simply doubled in the case of *Muga*.[1] The phenomenal increase in the production of *Pat* has been attributed to an active Research and Development work in the sector, an aspect which has been sadly neglected in the case of *Eri* and *Muga*.[2]

Of the three varieties, *Muga* silk has the greatest economic potential. Apart from the fact that it is the highest paying silk, Assam can fruitfully exploit the monopoly that she now enjoys to push the product in the national and international market. *Eri* also has tremendous prospects as dress and suit material if blended with wool or cotton. We have already noted how in the late nineteenth century this product was largely in demand in Europe and Australia. Recent experiments on *Eri* at the Central Silk Technological Research Institute of the Central Silk Board have resulted in the production of a high quality lustrous yarn which can be effectively blended with other silks and natural fibres. Thus, *Eri,* also has a lot of potential. In fact, the Central Silk Board is now planning to start production of *Eri* in Andhra Pradesh, Jharkand and Bihar.[3] Although Assam is the fifth largest producer of mulberry silk in India, *Pat* silk, has to face competition not only within India but from outside as well. Moreover, an influx of the cheaper Chinese silk has resulted in a steady decline in the demand for mulberry silk in the market.

In the circumstances, *Muga* is now being projected as the 'silk of the future'. In 2001-02, the production of *Muga* in Assam was 96 metric tonnes, spread over an area of over 3,000 acres. The projection for the Tenth Plan is to increase the area under cultivation to more than 4,000 acres and production to 150 metric tonnes. The Plan also seeks to provide assistance to *Muga* rearers for maintenance of plantations and *Muga* seed multiplication infrastructure apart from support in the form of crop insurance. Today, *Muga* silk has been identified as one of the most potential sectors in Assam capable of earning foreign exchange. Ironically, in spite of this realisation, the industry is plagued with problems, most of which can be easily resolved provided there is a will to do so.

Current Problems Faced by the Industry

As a result of our enquiries[4] we have been able to identify the main problems currently faced by those engaged in the industry.

- Apathetic attitude of the Government.
- Lack of basic necessities like adequate power and water supply.
- Obsolete production methods.
- Difficulties in procurement of raw material.
- Dominance of middlemen.
- Lack of funds.
- Lifting of subsidy on plant and machinery.
- Inadequate Research and Development.

The apathetic attitude of the government has been primarily held responsible for the present state of sericulture in Assam. It is generally accepted that government initiative and control are very necessary to enable an industry to stand on its own feet and that the theory of *laissez faire* is not applicable at all times. Private enterprise no doubt needs to be encouraged but the government must be prepared to lend support by way of providing the infrastructure. In Assam, Government silk farms do exist, but most of them do not function efficiently. The Hiloibam Silk Farm near Dibrugarh is an example. Set up in 1981 on a 70 *bigha* plot, the farm was expected to produce 20,00,000 silk cocoons annually. Initially, under dynamic leadership and effective supervision, the farm functioned perfectly. However, within a few years, production fell and today it is in shambles. Ironically, the staff that had been appointed at the farm are still there. They have no work but draw salaries anyway. On the other hand, with a little initiative, another government farm, the Tengapukhuri Government Farm, is flourishing. The Assam Branch of the Indian Tea Association (ABITA), through its *Gramin Krishi Unayan Prakalpa*, has been active in bringing technology to farmers. One of its thrust areas is the *Muga* industry in the Tengapukhuri and Sapekhati area of Sivasagar District. At the government farms here, ABITA has provided reeling cum twisting machines at a subsidized cost of Rs. 5,000. This electrically powered machine can reel 200 grams thread or 1,000 cocoons in one hour. The farm has also been provided with machinery to spin thread from the waste products of cocoons.

In most sericulture farms, basic necessities like adequate power and water supply are missing. Government subsidy on plants and machinery was also lifted in 1987. Moreover, in most cases, the obsolete production methods do not make farms a commercially viable proposition. It is, therefore, obvious that production is not sufficient to meet the demand. In the circumstances, procurement of raw material is not easy and in the absence of any controlling authority, suppliers charge arbitrarily. Very often these suppliers are not the actual producers of the silk but middlemen who have entered the market as traders.

The State Government has displayed little initiative in helping silk producers to organise themselves into an economically viable industry. It has failed to pump in funds for Research and Development and to provide financial assistance to the producers to enable them to undertake the modernisation and up-gradation required to retain a premier market slot. This is in stark contrast to the initiative taken by the Government of West Bengal which is trying to cash in on the popularity of Assamese silk by drawing up a long-term scheme to produce *Muga* in the state. It has recently formulated a plan involving around Rs. 300 crores to promote *Muga* in Coochbehar. The Government plans to provide Research and Development support to the industry through a cell set up at the North Bengal University. As long as there was little competition the silk manufacturers of Assam could somehow carry on despite the drawbacks and administrative apathy. But with West Bengal set to give competition this indigenous industry of Assam has a bleak future if the State Government does not rush to the aid of the beleaguered producers.

Recommendations

Sericulture consists of two activities: plantations and rearing. So, in order to increase production, both these aspects have to be looked into.

- **The increased production of leaves**, on which the silkworms feed, is the key to increasing silk production and therefore is of utmost importance. This can be achieved by
  (i) Increasing land area under plantation
  (ii) Increasing leaf production per unit area. This is possible if superior quality plants are used and improved plantation and harvesting techniques are adopted.
- **The production of silkworms can be raised** by

(i) **Using good quality eggs.** The development and sustenance of sericulture depends on the use of good quality eggs, scientifically termed as "disease-free layings." The process of production involves special care in breeding and maintenance of improved varieties of silkworm. Hence this aspect should not be neglected. The supply of good quality eggs should be ensured.

(ii) **Adoption of improved rearing methods.** A thatched house with mud walls is suitable for rearing silkworms. The house should be well ventilated and free from water logging, dampness or direct sunlight. Suitable temperature should be maintained by artificial heating/cooling as required.

These details, which are available with the State Sericulture Department and Central Silk Board, should be given adequate publicity among the producers.

- **Research and Development** is extremely important for increased production. We have observed earlier that the production of *Pat* is much higher than *Eri* or *Muga* and that this has been attributed to greater R and D work in the field. Now, Research and Development in respect of *Muga* and *Eri* need to be strengthened with the following objectives:

(i) developing improved varieties of food plant
(ii) optimising leaf production
(iii) evolving improved qualities of silkworm
(iv) improving rearing techniques

Other suggestions include the following:

- **Assistance** in the form of loans, subsidized machinery, seeds etc
- **Technical Information** to be provided to the producers
- **Modern machinery** to replace the obsolete ones now in use
- **Technical Training** to be imparted
- **Government control,** at least initially, over the price of raw material ought to be exercised
- **Uninterrupted supply of basic necessities like water and power.**

The age old silk industry of Assam has won for her a place in the economic map of the world. Today, in spite of stiff international competition, it is still possible to hold on to that position. But this will be possible only with active Government support and encouragement. To keep pace with the rest of the world, the entire industry—from plantations to rearing of silkworms to the manufacture of silk—needs to be thoroughly overhauled.

References

1. Information available from the *Department of Sericulture, Government of Assam*, Guwahati.
2. *Ibid.*
3. *The Assam Tribune,* 12 October, 2002.
4. Information collected from interviews with craftsmen and others engaged in the industry.

HANDLOOM AND TEXTILE INDUSTRY

Closely associated with sericulture is the handloom and textile industry in silk and cotton. In fact both go hand in hand and one is dependent on the other. We have seen that since ages weaving has been an integral part of Assamese society and that for long the handloom has been an essential part of every Assamese household. With increasing competition from cheaper mill-made cloth, however, handlooms now supply only certain types of cloths. Even in rural areas most of the cloth is woven from mill-made yarn.

One significant aspect of the textile industry in Assam is that the products are varied, colourful and vibrant. The different ethnic groups that dwell in this region have their own designs that distinguish them from those of the other groups. The fabrics, woven either in silk or cotton are embroidered in the loom with gold or silver wires or in coloured yarn. It is also very common for the weavers to blend different material to produce a fabric.

The Government of Assam, in an effort to showcase the handloom products of Assam, has set up ARTFED, a co-operative federation for producing handloom and handicraft products of the region. Apart from setting up emporia in different places, ARTFED also organises Handloom exhibitions from time to time, within the state and outside, in order to promote its products. Very recently ARTFED sent a consignment of *Muga* duvets worth $ 18,732 as part of the total consignment worth $ 75,000 ordered by the American chain store, Bloomingdales. The export turnover of ARTFED till December 2003 was Rs. 2.98 crores and is expected to increase to Rs. 5 crores by the end of the financial year.

The North-Eastern region of India accounts for 53 per cent of the looms and 50 per cent of the weavers in the country, but they account for only 20 per cent of the total production with only 13 per cent share in commercial looms. The handloom sector provides employment to thousands of people but owing to low productivity, the economic benefits are not proportional. Nowadays, handloom products have become highly fashionable in the global market and the demand for it has increased tremendously in the last few years. Assam, with her strong background of the weaving industry, should have been able to compete nationally if not internationally. Growth and development, however, have been limited.

Problems Faced by the Sector

The textile industry in Assam is currently facing a number of problems. Unless these are attended to, the industry cannot be expected to emerge as a competitor in the national and international field. We were able to identify the following main difficulties faced by the weavers.

- Size of the unit.
- Finance.

- Marketing.
- Technology.
- Non availability of raw material.
- Power shortage.
- Lack of proper transport facilities.

The size of the unit is generally very small, both in terms of the number employed and the volume of output. One of the reasons for their small size is their household character. The major motivation in operating the unit is to earn livelihood for the family rather than setting it up as a commercial proposition. Even at Sualkuchi, where more than 75,000 people including around 30,500 weavers are directly involved in the industry, most of the units have only around 7 to 10 looms. There are, of course, exceptions where some factories have as many as 150 looms with around 400 employees. Production, therefore, varies according to the size of the units and in most cases, after meeting the local demand, there is little left for sale outside.

Absence of finance is a major constraint in the growth of this sector. A serious problem of rural industries is in respect of credit—both long-term and short-term. The credit situation is particularly hard for the very small units. Most of the weavers complained of the following finance related problems:

(a) Unavailability of timely credit facilities,
(b) Tedious process involved in getting a loan,
(c) Complicated loan application form,
(d) Too much insistence on collateral securities, and
(e) Harassment by bank officials in a number of cases.

As a result many are compelled to take recourse to the private money lenders who charge high rates of interest.

Marketing appears to be another chronic problem. In the backdrop of mass production from outside the state overwhelming the cottage industries of Assam, the absence of administrative support for certain traditional items having the capacity to find regional, national and even global acceptance is disheartening. The *Gamocha* (soft cotton towel) may be cited as an example. Given its utility and relative cheapness, a sizeable market for this product exists not only within the state but outside as well. But a virtual deluge of cheaper power-loom *Gamochas* from southern states like Andhra Pradesh had, till recently, offered such a stiff competition that the local industry was almost wiped out. Fortunately, the State Government's decision, though belated, to impose a ban on the sale of imported *Gamochas* has helped to put the industry back on its rails.

Marketing requires considerable support from various agencies. In the absence of a proper advertising channel, or an agency to conduct market surveys and analysis, the weavers are left to fend for themselves. Moreover, wholesalers and retailers often accept goods on long term credit only. This stifles the flexibility of the producers and constricts the availability of money in their hands so essential for circulation. The high rate of sales tax is of course another deterrent.

Like the sericulture industry, here also up-gradation and modernisation of machinery 6 is of utmost importance. The majority of the weavers have no access to modern training and technology and automatic looms. They adhere to outmoded techniques which obviously accounts for the slow and limited production. Hence it is extremely difficult to keep up with international standards. Another major problem is the difficulty that the weavers face in the procurement of raw material. As demand exceeds supply, middlemen have entered the field and, in the absence of a controlling authority, they impose rates arbitrarily. This increases the cost of production and automatically places them in an adverse position.

The apathy of the government has been cited as a major deterrent once again. It has not been able to provide basic infrastructural facilities like uninterrupted power supply or an effective transport system in the rural areas and these have largely hampered the growth and development of the industry. The much publicised Golden Thread Project, set up in Sualkuchi by the State Handloom and Textile Department, which has remained non-functional since its inception in September 2000, is a glaring example of government apathy.

Recommendations

The handloom industry of Assam has tremendous prospects. But if a flourishing industry is to be established, the following points need to be addressed.

- Use of modern machinery.
- Adaptation of designs to meet national and international requirements.
- Diversification of styles and production of fabrics in blended yarn.

- Adequate publicity.
- Proper marketing.
- Setting up co-operatives.
- Investment of capital.
- Government support.

The use of modern machinery is of utmost importance. Improved production techniques will ensure the production of high quality material which is essential if the industry is to survive in today's competitive world. Moreover, supply must be commensurate with the increasing demand and this will only be possible with mechanisation.

The products of our handlooms evoke images only of the *Gamocha* and the traditional attire of the Assamese—the *mekhela, chador, riha, eri chador, dokhona etc.* Unfortunately, the products of our handloom industry have not yet moved muc beyond these items. A few young designers have tried to bring about certain innovations in the fabrics, but they are still in the xperimental stage. What is now required is a big push for the take off. Diversification of styles and a variety in the production of fabrics in blended yarn is essential if the products are to be marketed globally. Moreover, apart from dresses and dress material, other products such as linen and upholstery material should also be borne in mind.

A primary feature of a p l ny industry is investment capital and the availability of credit facilities is a critical and catalytic element. Here again the State has to take the initiative in granting financial aid to the craftsmen. The bottlenecks that hinder smooth credit flow need to be identified and removed as far as possible so as 7 to nable weavers to take advantage of the various schemes offered by the financial institutions. Very often the craftsmen are not even aware of the programmes that exist. Hence, an awareness programme needs to be conducted urgently.

The handloom industry of Assam has tremendous economic potential. But along with intensified production in a variety of products, effective publicity and marketing are equally important. In order to improve technology and to provided proper marketing backup to the thousands of weavers, co-operatives are a must. In Assam, less than 30 per cent weavers are members of co-operatives.

CANE AND BAMBOO

We have observed earlier that from time immemorial, cane and bamboo have formed an integral part of the lives of the people of North East India. As such the people of the region are important stakeholders and repositories of knowledge which can be extremely useful for the development of this sector. Bamboo and cane are both such versatile material that they have survived in spite of competition from other products. Bamboo particularly is not only a viable alternative to wood, but is also valuable as a natural insecticide and pesticide. Further, it has a variety of medicinal properties suitable for curing a number of skin ailments and diseases like diabetes. Bamboo and cane furniture and handicrafts are produced throughout the region, but because of limited knowledge, lack of skills and basic technology, productivity is very low. Moreover, the quality of the products is also generally poor. Very often, the bamboo that is used is not adequately mature and because of the limited expertise of the craftsmen regarding jigs and jointing, the finishing is not up to the mark. In the circumstances the necessity of a specialised institution dealing solely with bamboo and cane was strongly felt and it was in this context that the Cane and Bamboo Technology Centre (CBTC) was conceptualised.

The Cane and Bamboo Technology Centre was set up in Guwahati with the object of identifying and disseminating knowledge among the craftsmen in this field and to network with specialised institutions within India and abroad. The United Nations Development Programme has committed approximately $ 1.4 million for technological up-gradation and networking of the Cane and Bamboo Project. While the Department of Science and Technology, Government of India, is the executing agency for the project, the implementation agency in Assam is the North Eastern Development Finance Corporation Ltd. (NEDFI).

The CBTC intends to promote employment and income generation by making relevant technologies and business opportunities more accessible. In an effort to upgrade the skills of entrepreneurs and craftsmen, certain resource centres have been identified for the purpose of imparting training. It is also hoped that a long term association will be established with the International Network of Bamboo and Rattan (INBAR), based in Beijing.

The importance of bamboo can be gauged from the following facts:

- It is the fastest growing plant. It grows three times faster than the fastest growing tree. Some species grow at the rate of 1 metre per day. This growth pattern makes it easily accessible in the minimal amount of time.
- It is a viable alternative to wood. Bamboo is the strongest building material. Its tensile strength is 28,000 per sq. in. against 23,000 of steel. Bamboo timber has the additional advantage of being made

from an abundantly available raw material. Considering the increasing environmental consciousness regarding deforestation, the market for bamboo is tremendous.

- It is eco friendly. It generates more oxygen and enriches the soil.
- It is an enduring natural resource and can be harvested selectively annually.
- It is a versatile raw material with a short growth cycle. There are over 1,200 species of bamboo. This makes bamboo adaptable to many environments. It can be harvested in 3 to 5 years time as against 10 to 20 years in the case of most soft woods. Moreover it tolerates extreme precipitation from 30 in. to 250 in., annual rainfall.
- Bamboo is directly related to the economy of a region because it provides income, food and housing to 2.3 billion people worldwide.
- It is flexible and lightweight and is thus suitable as building material for earthquake-prone regions.
- It possesses a variety of medicinal properties.
- It is a valuable source of high quality carbon.

Current Problems

In spite of all these positive aspects, not much headway has been made in utilizing these concepts commercially. Although the uses of bamboo are so varied, its use in Assam is limited to handicrafts and furniture only.[1] Several groups of artisans/micro entrepreneurs have developed. The most important are in the Lower Brahmaputra Valley (Nalbari, Barpeta Nagaon), Upper Assam (Lakhimpur), and Barak Valley (Silchar, Karimganj). They produce furniture, household and decorative items. Production meets local demand although a certain quantity is also marketed outside the state. Unfortunately, the export market is almost non-existent. This has been the case for a variety of factors, some of which have been identified below:

- Lack of exposure of the artisans.
- Obsolete technical know-how.
- Non-availability of proper and cost effective technology.
- Non-availability of proper indigenous machinery.
- Crude finishing.
- Limited use.
- Non-availability of technology for the utilisation of waste economically.
- Absence of a centralised quality control agency.
- Exodus of raw material from the region.

The government has initiated certain steps in an effort to solve the above problems but they have not proved adequate. In the absence of a quality control agency, products are very often manufactured from immature bamboo and poorly finished. The nailing process, for instance, is so crude that the products are not likely to be accepted at all in the international market. Moreover, the limited use of cane and bamboo in Assam have resulted in an exodus of the raw material to other regions. Hence, artisans often face difficulties in procuring raw material.

Recommendations

Bamboo has tremendous economic potential. Like *Muga,* Bamboo, too, can provide livelihood to thousands of people in Assam and be a major foreign exchange earner for the region. China's business in bamboo is over one billion dollars. This has been possible because of the high quality and cost effectiveness of her products. Hence it is of utmost importance to ensure that the products are of an international standard.

The following products are likely to have an international market:

- **Bamboo floorings, panels, furniture wood, blinds, beams and columns.**
- **Bamboo vinegar** which is internationally used in insecticides and pesticides, as a deodorant in agriculture, medicine, cosmetics, food processing, sewage disposal etc.
- **Bamboo charcoal** which has its use in the metallurgical industry, carbide battery and for various grades of carbon in the form of activated carbon.
- **Bamboo oil** for use in medicines and hair and skin care products like shampoos and ointments.

It is significant to note that a number of by products like bamboo briquettes can be manufactured from the waste that is generated from the above mentioned products which are just a few possible industries that can be set up in Assam. A bamboo mat corrugated sheet unit has been set up at

Byrnihat, near Guwahati. The sheets are light but strong and have shown very good resistance to water, fire, termites etc. The plant has a production capacity of around 3,500 sheets per month.

Most people are unaware of the tremendous potential of bamboo. Therefore, an awareness campaign is essential to educate the people on the subject and to bring out entrepreneurs to the field.

References

1. The utilisation of bamboo in the organised sector, viz., production of paper and pulp, has been left out of the purview of this study.

HORTICULTURE

Assam has unique agro climatic conditions which is conducive to the production of a variety of horticultural crops. Agriculture and Horticulture play a crucial role in the state's economy with 75 per cent of population deriving their livelihood from this sector. About 90 per cent of the farmers belong to small and marginal groups, the average operational holding being 1.37 hectares. Horticulture encompasses a wide range of fruits, vegetables, flowers and ornamental plants, aromatic and medicinal plants. The total area under horticultural crops is almost 5,500,000 hectares which produces 13,00,000 tonnes of fruit and more than double that quantity of vegetables annually, besides spices, nuts, and flowers. The homestead garden, locally known as *Bari,* is a unique feature of rural Assam as it generally meets the day to day requirements of the farmers. At present the productivity of most of the horticultural crops in Assam (except oranges) is below the national average. For instance, the production of bananas is 13 tonnes per hectare against the national average of 27 tonnes per hectare. Apart from a variety of common fruit like oranges,, papayas, jackfruit, guava, mango and litchi, which are found in plenty, the region abounds in a number of other fruit like *amlakhi* (*Phyllanthes emblica*), wood apple or *bel* (*Aegle mamelos*), *leteku* (*Baccurea sapida*), *amora* (*Spondius mangifera*), *kathal* (*Artocarpas heterophyllus*), *kuji thekera* (*Garcimia cowa roxb*), fig or *dimoru* (*Ficum carica*), *poniol* (*Flacourtia gangomos*), *silikha* (*Terminalia chebula*), olive or *jolphai* (*Elaeoearpus floribundus*) etc. most of which have a lot of medicinal value. A few with medicinal properties are listed below:[1]

Bel: An important indigenous fruit of India, the *bel* tree grows wild in Assam. The ripe fruit is a laxative while the unripe fruit is prescribed for diarrhoea and dysentery. *Bel* is an extremely nutritious fruit and 100 gms of its pulp contains 61.5 gms. of water, 1.08 gms. of protein, 0.39 gms. of fats, 1.70 gms. of minerals, 31.8 gms. of carbohydrates, 55 mg. of carotene, 0.13 gms. of thiamine, 1.19 gms. riboflavin, 1.1 mg. of niacin and 8 gms. of Vitamin C. No other fruit has such high content of riboflavin. The ripe fruit also provides relief for rheumatism.

Amlakhi: Amlakhi is another fruit which usually grows wild but is also cultivated. The fruit is highly nutritive and rich in Vitamin E (440–600 mg/100 gm). The fruit can be used to make preserves, sauce, candy, dried chips, jellies, pickles etc. It is valued as a diuretic alternative and antibiotic and is used in the treatment of chronic dysentery, diarrhoea, jaundice, dyspepsia, diabetes and cough etc. It is also used in the tanning and dyeing industries.

Amora: This fruit is rich in vitamins. It purifies the blood and is effective in the treatment of scurvy, rickets, bacillary dysentery etc.

Kathal: Assam is the highest producer of jackfruit in India. It is largest amongst edible fruits. The tender fruit and the seeds can be used as vegetable while the ripe fruits can be eaten raw. The fruit is rich in protein, acids, vitamins and minerals. The waste product, such as skin and core of the fruit, can be used commercially for pectin extraction.

Kuji Thekera: It grows wild in Assam although it is sometimes also cultivated in the backyards of Assamese households. The fruit is highly acidic and usually preserved as sun-dried slices and very commonly used in preparation of dishes. It is widely used against dysentery. Its Vitamin C content is 4.05 mg./100 gm.

Dimoru: Fig trees grow wild in Assam and are not generally cultivated. The fruit, both fresh and dry, is valued for its laxative property. It also has other medicinal properties and is used in the treatment of boils and other skin infections. Every 100 gm. of the edible portion of *dimoru* contains 4 gm. protein, 200 mg. Calcium, 4 mg. Iron and Vitamin A.

Nemu tenga: (*Citrus lemon*) The lemon plant is an evergreen plant which is actually a shrub. It is a major cultivated fruit crop of Assam occupying an area of 8,700 hectares. The fruit is oval, unlike other lemons which are normally round, and is very juicy. It contains a high percentage of Vitamin C (10 mg.) besides Vitamin B (0.02 mg.) and Niacin (0.1) per 100 gm. Because of its high Vitamin C content, it is effective in the treatment of skin diseases and can give relief to pain in joints, coughs and colds. Lemon juice also helps to control high blood pressure.

Poniol: This fruit is a rich source of protein (3.65%), Vitamin C (217.99 mg./100 gm.) and minerals such as Phosphorous (146.80 mg./100 gm.), Calcium (175.50

mg./100 gm.), Potassium (158.10 mg./100 gm.) and Iron (118.30 mg./100 gm.). The iron content of Flacourtia is 280 times more than apples. It also contains several essential amino acids.

Jolphai: This fruit is often grown in backyards of Assamese households and is also available in considerable quantities in the wild. It is a good source of Calcium, Phosphorous, Iron, Vitamin B and Niacin. Olive oil is valued for its medicinal properties and its use as a cooking medium.

Silikha: Silikha, the fruit of a tall tree that grows wild in the State, is valued for its high medicinal properties. It is a small, oval fruit which is green when fresh and black and hard when dry. It has a wide range of medicinal uses and is a major component of many Ayurvedic medicines. It helps to cure indigestion and is reported to be beneficial in the treatment of a number of ailments like Asthma, Piles, colic pains, heart diseases, Scabies, Jaundice etc.

Mushroom: Mushroom, an important protein food acclaimed all over the world, is a favourite item in the diet especially of the tribal people of the State. The production of mushroom is gaining popularity in the State for its nutritive and commercial value. Edible mushroom contains 47.93 per cent of protein, 4.10 per cent of fat and 9.5 per cent of minerals and is comparable to other sources of protein like meat, eggs, fish, milk, soya bean and pulses. It contains high quality amino acids. Owing to its low fat and carbohydrate content, mushrooms are an ideal protein food for heart and diabetic patients. Dried mushrooms have a long storage period.

Ginger: Ginger is a major spice crop of Assam occupying an area of almost 16,309 hectares. Apart from its use as a spice, ginger, which has a wide range of medicinal properties, is used in the treatment of ailments such as colic pain, coughs, asthma, influenza, etc.

Jaluk (*Piper nigrum*): *Jaluk or* black pepper, known as king of spices, is also called Black Gold. Of late, this perennial creeper is gaining popularity as a commercial crop and at present the cultivated area of pepper in Assam is around 2100 hectares. The crop is highly valued for its content of Oleoresin, and essential oils.

Tamol (Areca catechu Linn): Arecanut is a traditional major cash crop of Assam cultivated in an area of around 74,500 hectares. It is widely used as a condiment with betel leaf. Besides, it is used in the preparation of dyes. Dried and processed *Tamol* are converted into *supari*. *Tamol*, also known as betel nut, is an appetizer and is frequently used as a remedy for secretion of bile and cough. An important by product of the betel nut tree is the fibre, known colloquially as *dhokua*, that is obtained from the stem of the leaves. In Kerela a flourishing industry exists for the manufacture of disposable plates, bowls etc. from this fibre. As the material is biodegradable, it is environment friendly and is likely to find an international market.

From the above it is apparent that Assam has tremendous potential not only for a fruit processing industry but also for an industry in indigenous medicine. Till now, however, not much headway has been made in exploiting the resources in this sector.

Problems Faced by the Industry

Certain constraints have impeded the development of horticulture in Assam. Some of the problems that have been identified are listed below:

- Non-Availability of Quality Planting Material.
- Lack of Awareness of the Potential of Horticulture as a Commercial Crop.
- Weak Marketing Structure and Management.
- Lack of training Facilities.
- Inadequate Technology.
- Absence of Cold Storage Facilities.
- Scattered Production.
- Difficulties of Transportation and Communication.

The availability of quality planting material plays an important role in the development of Horticulture. In Assam, the present system of purchase of the seedlings, layers, grafts etc. from private nurseries by the farmers is a risky proposition as there is no regulation to guarantee the quality of the planting material and it is seen that often these traders mislead the farmers.

There is also very little awareness among the farmers in general about the commercial prospects of horticulture. Moreover, the marketing infrastructure in the state is very traditional with the interference of a long chain of intermediaries. Further, there is no facility in any market for assembling, grading, storage or display of the produce. Owing to the difficulties in marketing the produce, most of the farmers resort to either pre-harvest sales or the immediate post-harvest sales thereby creating a glut in the market. In the absence of medium or even small scale processing centres, this problem is aggravated by the additional fact that there are hardly any cold storage facilities in Assam.

Another important constraint is that production is very scattered and hence, in the absence of an organised market and difficulties in transport and communication, it is often difficult to bring the produce to the consumer.

Recommendations

Assam's agro climatic conditions offers huge potential for a wide range of horticultural products which can be commercially exploited. Currently organisational support is available from the Assam Agricultural University and the Indian Council for Agricultural Research. Moreover, there are certain Field Trial Stations and Regional Research Stations. State intervention has been in the form of the Technology Mission for Integrated Development of Horticulture and training and demonstrations by the Assam Rural Infrastructural and Agricultural Sub-Project (ARIASP) and National Horticulture Board. Apparently, the support is not adequate and much more needs to be done if Assam is to emerge as a forerunner in horticultural products. The following points need to be considered.

- Emphasis should be on the production of low volume, high value, less perishable goods.
- Development of Nurseries with good quality mother plants.

- Development of a market infrastructure.
- Development in the communication and transportation system.
- Creation of post harvest facilities.
- Thrust crops like cashewnuts, betel nuts, bananas, medicinal plants, pepper etc. need to be identified.
- Development of processing centres.
- Mechanization.
- Creation of an awareness among youth of the commercial viability of horticultural crops in the national and international market.

Assam's potential in the horticulture sector is immense. But these require organised and sustained efforts to establish and promote. But most significantly, the rural producers must be able to get adequate returns for their toil. Falling prices owing to abundant produce that cannot be transported to the consumer market in a timely manner is a major drawback. Further, the transportation and other facilities provided for should be affordable. Cold storages may be set up but they will be of little help if the farmer cannot afford to use them. The government should also ensure adequate and non-intermittent power supply so that the industry can be mechanised. Government help is of utmost importance. At the initial stages, the producers must be guaranteed by the government a fair return on their toil and investments. Most of the extension programmes that exist are not effective. They must be overhauled to provide training in improved methods and techniques. Finally, stress ought to be laid on Research and Development. For instance, the indigenous fruit and plants with high value medicinal properties can be identified for making medicine. With the global trend leaning towards organic food, the indigenous medicine of Assam is bound to have a world market.

The horticulture sector in Assam thus holds immense possibilities. What is now required is modernisation in production, processing, and marketing. With encouragement and initiative from the Government it is hoped that more and more entrepreneurs and Self Help Groups will come forward to involve themselves in these employment and income generating projects.

References

1. Information collected from ICAR Complex for NEH Region, (Division of Horticulture), Umium, Meghalaya.

FLORICULTURE

Closely allied to horticulture is floriculture an industry which in recent years has scaled new heights in India. There are now in India over 500 export oriented units in operation[1] with over 200 hectares under cultivation for the production of cut flowers. The North East is very rich in flora with large numbers of ornamental and flowering species growing in wild and semi wild conditions. In fact, 824 species of orchids alone have been identified in this region of which many are available in Assam. The most popular orchid of Assam is the *Kopouphul.* Besides orchids, the state abounds in a variety of exotic flowering and non-flowering plants which may be commercially cultivated. Considering its tremendous potential, floriculture in Assam is often referred to as the "Sleeping Giant".

In recent years, there has been a growing demand for flowers, especially in the urban areas. This has resulted in the emergence of hundreds of nurseries all over the state. But the cultivation of flowers on a commercial scale is yet to take off. In an around Guwahati there are a few farms which are devoted solely to floriculture but they are limited in number. An orchid farm set up by India Carbon has a thriving export business. With a view to promote exports of floriculture products from North East Region, a "Buyer Seller Meet" was organised at Guwahati in June last year[2]. Encouraged by the response and enthusiasm of potential entrepreneurs during the meet, a small workshop on "Commercialising Floriculture in North Eastern Region" was also subsequently organized where an 'Action plan for Commercial Floriculture Development in North Eastern Region' was prepared and circulated among the organizations associated with the programme such as NEDFI, NEC, National Horticulture Board (NHB), NABARD and State Directorates of Horticulture etc. It was felt that at present the region lacked the infrastructure necessary for facilitating successful commercial enterprise in floriculture. It was also observed that the dissemination of knowledge regarding post-harvest handling in the form of appropriate technologies, among growers was essential to check deterioration of flower quality and improve profitability of farmers. Hence suggestions were made to impart training in these respects to prospective entrepreneurs.

The Floriculture industry in Assam has unlimited possibilities. The climate is highly favourable for the cultivation of a number of flowers like, marigold, gladiolus, gerbera, anthurium, chrysanthemum etc., apart from orchids, all of which have an assured global market. A concerted effort should, therefore, be made to develop this industry on a commercial scale. But apart from cut flowers, there is also scope for an industry in floral handicrafts made out of dehydrated flowers and foliage. The process of drying the flowers and foliage is simple and does not require much capital investment. On the other hand there is a considerable demand in the national and international market for pot pourie, greeting cards, calendars, wall hangings, framed designs etc. made out of pressed flowers. Hence the production of floral crafts can be a remunerative cottage industry.

Handicraft

Assam enjoys a place of pride in the country with its rich heritage of artistic craftsmanship. The wide range of handicrafts are almost entirely made from locally available raw material. The principal items include basketwork, cane furniture, mats woodcarvings, terracotta, artistic textiles, bell metal work, brassware, dolls and handloom products like mats and wall hangings. Kamrup, Sivasagar and Nagaon are best known for basketwork and mats. The Karimganj sub-division of Cachar district and Bokajan area of Karbi Anglong are famous for a very special type of mat known as *Sital patis.* Hajo, Sarthebari and Baligaon, the most flourishing centres for the bell metal industry in the nineteenth and early twentieth centuries, still produce these items though in a much more limited scale. Similarly, the *Japi* (traditional headgear of the Assamese), survives despite many handicaps.

Problems Faced by the Industry

The industry in handicrafts has not been able to develop professionally in Assam in spite of local talents, skill and abundance of raw material. As a result, many of the traditional items are fast disappearing. This is indeed unfortunate. The handicraft sector is economically important because of low capital investment, high value addition, negligible import content and high potential for export earnings. In recent years, the handicrafts sector in India has seen vast expansion in terms of the national and international market and has proved to be an important source for foreign exchange earnings. In fact, it is the largest employer of manpower after agriculture in the rural sector. In Assam, however, the handicraft sector is struggling to survive. In our study, we were able to identify the following reasons.

- Unorganised production base and inadequate marketing network.
- Lack of modern technology.
- Scarcity of raw materials.
- Limited credit facilities.
- High production costs.
- Stiff competition.
- Lack of exposure.
- Government apathy.
- Exploitation by middlemen.

The handicraft sector in Assam does not have an organised production base. Nor is there an effective marketing network. Although a number of emporia and promotional organisations like the North Eastern Handloom and Handicraft Development Corporation, ARTFED etc. have been set up, these institutions are yet to play a significant role in enhancing the earnings of the artisans. Very often, the main channel for marketing the artisans' produce is the network of middlemen and traders, thereby robbing the artisan of his share in the huge profits that are made.

As in the other industries, the handicraft industry also suffers from the lack of modernisation of technology. Most of the craftsmen still continue the age old methods of production which are crude, time consuming and not at all cost effective. As a result, the products are expensive and at the same time not up to the mark in quality. The bell metal products may be cited as an example. The *Sarai, Bota, Dunari, Kaloh, Lota, Kahi, Bati,* to name just a few, are all beautiful traditional Assamese household items, but the finishing is so poor that in their present form there is no scope for an export market. Hence consumption is limited to the local market.

Apart from lack of mechanisation and application of modern technology, artisans working in cane and bamboo suffer from additional drawbacks. Cane, which was once abundantly available in the forests of Assam, has now become scarce. Apparently there has been no concerted effort by the Forest Department to regenerate cane by introducing its cultivation on scientific lines as has been done in Malaysia and Indonesia. Moreover, in the absence of any restrictions on the export of cane to places outside the region, much of the raw material that is available is siphoned out. As a result the price of cane has increased manifold in recent times. Almost 55 per cent of the handicraft units in Assam deal exclusively in cane. The shortage of raw material together with the high prices, have driven many artisans in this field to extreme poverty.

Competition in the handicraft sector is very stiff. In states like Jammu and Kashmir, it is a major revenue and foreign exchange earner. In fact, by 2009, the handicraft exports from that state is expected to touch Rs. 1,500 crore. The government has undoubtedly played a significant role in boosting the handicraft industry. In Assam, this appears to be sadly lacking.

Recommendations

In order to solve the above problems certain steps need to be taken.

- **Improvement in the supply of raw material.** This is especially applicable to the cane and bamboo industry. As Assam is ideally suited for the cultivation of these two products, the State Government should introduce large-scale cultivation of cane and bamboo in the forest areas. Moreover, people can be motivated to grow them as cash crops. The government can lend support to this sector by providing scientific training to the prospective cultivators and by granting certain subsidies.
- **Mechanisation** is an extremely important factor. This would mean quicker production of greater numbers, which would result in lower costs. Machine made products are also likely to have a better finish. This would enable the products to compete nationally and internationally.
- The **designs** of many items need to be modified or changed to meet the requirements of the global market. However, it is important to remember that while evolving new designs care has to be taken to retain the traditional designs as far as possible. Most of our existing handicrafts, for instance, are rather bulky which makes them difficult to transport. A change in design might solve the problem.

  Nowadays, vegetable dyes are very much in fashion. Assam has a rich tradition of these dyes which can be improved upon and used fruitfully.

- **Credit Facilities** is another important area that has to be looked at in depth. The majority of the artisans have very little awareness of the various incentives and promotional schemes that are offered by the funding agencies. Aggressive public campaigning to educate the people in this regard is urgently needed.
- **Improvement in the marketing network.** Publicity is a very important factor in attracting potential customers. All efforts should be made to promote the indigenous products, through exhibitions, seminars, workshops, stalls in various airports, railway stations bus stations, commercial centres etc.
- **Improvement in infrastructural development** like transport and communication
- **Government support** is essential if this sector it is to flourish.
- **Exposure.** The majority of our artisans lack exposure. Study tours, visits to different handicraft fairs, handicraft centres etc. in different states may be organised so as to expose them to new processes, new designs, new techniques and new ideas in general.

Assam has all the potential for a vibrant handicraft industry: manpower, skills, expertise and natural resources. What is now essential is government support and help and co-operation from the concerned departments, financial institutions, professionals and organisations. It is heartening to know that Indian corporate giants are planning to include traditional handloom and handicrafts of Assam in their list of gifts. The Confederation of Indian

Industries has recently taken up a special agenda for catalysing the development of an effective marketing system for the crafts of this region through the leadership of the corporate sector. This will no doubt give a definite boost to the handicraft industry of Assam.

References

1. P. Nageswari, Globalisation of Village Industries Products, *in* M. Soundaripandian, ed., *Industrial Resurgence in Rural India,* New Delhi, 2000, p. 53.
2. *Assam Tribune,* 11 June, 2003.

MASKS

Closely allied to the handicraft industry is the manufacture of masks in Assam. We have observed that in Assam masks were a significant mode of cultural expression and that they were widely used in folk theatre and in rituals by certain communities. Today, this form of folk theatre has almost disappeared, and together with it, masks have also become a rare sight. The few that are available, have been relegated to the status of wall decorations.

With changing times, modes of cultural expression are bound to change. The *jatra* and *bhaona* have made way for a more sophisticated form of theatre and the old props are no longer in demand. Gradually, the art of mask making has also declined. But fortunately the expertise required for making these masks has not completely disappeared and the craft is still practised by a few. In Thailand, a similar situation arose with the decline of the classical Thai theatre in which Khon masks were used. But there the industry in mask making was not allowed to die. Today, Khon masks are available in plenty in miniature or original, catering not only to the serious collector but also to the thousands of tourists that visit the country. In fact, the mask-making industry is an extremely flourishing industry in that country.

Assam, too, has the potential for a similar resurgence in mask-making. But currently the artisans face a number of problems which have to be removed if the craft is expected to grow and develop into an industry.

Problems Faced by the Craftsmen

- Expertise limited to a very small number of people.
- Production process extremely slow.
- Limited demand.
- Unorganised sector.
- Lack of marketing facilities.
- Infrastructural bottlenecks.

The above are some of the problems faced by the artisans. The manufacture of masks today is limited to the *Satras*. The expertise is handed down from father to son and no attempt whatsoever is made to train others in this age-old craft. Masks are not looked upon as a commercial proposition at all and are made to meet the occasional demand and in an effort to keep the tradition alive within the family.[1] The production process is also very slow and tedious. For instance, one person takes around 3 to 4 days to make a single face mask while it takes around one month to complete a big one. With so much labour involved, it obviously cannot be an economically viable industry in the present circumstances. Moreover, demand is so limited that it does not pay to practise it as a full time occupation. As a result, mask-making is now almost a dying art.

Recommendations

A concerted effort will have to be made if the industry is to be revived. The following points may be considered.

- Adequate publicity.
- Creation of awareness of its commercial potential.
- Creation of a market.
- Mechanisation of the manufacturing process.
- Organised production base.
- Training.
- Improvement in the marketing network.
- Improvement in the communication network.
- Government support.

The art of making masks is now almost totally confined to the *satras* of Majuli and to a few other places. Exposure to the craft is limited to the visitors who visit this river island usually between November and April and to occasional exhibitions held in different places. Adequate publicity therefore needs to be given to raise awareness not only about its shaky existence but also about its potential in the national and international market. In the absence of adequately skilled persons, training in the art will have to be imparted to prospective entrepreneurs. Moreover, the entire manufacturing process needs to be overhauled with the use of machinery so as to make the production quicker and cost effective. Supply, however, will only respond to demand and for this, a market has to be created. Further, it will have to be ensured that the finished products reach the consumers in time and for this an effective transportation and communication network is essential so as to link the producer to the customer.

All this, however, will not be possible without active government support. A sustained effort will have to be made to save the art from dying and to promote its products in the market in all possible ways.

Reference

1. Interview with Hem Chandra Goswami of Natun Samaguri Satva on 22.9.03.

MISCELLANEOUS

The industries discussed, are some that have immense prospects both in the national as well as in the international market. But apart from them, there are several other indigenous industries which also have plenty of potential. Mention may be made of the following:

Jewellery

We have observed that Assam has a variety of indigenous jewellery manually crafted by the goldsmiths. Unfortunately, today this is a dyeing art. In the face of stiff competition from machine crafted, lightweight gold and silver jewellery, the industry is unable to hold on to its former position. There is a dearth of skilled craftsmen for, like the mask makers, many goldsmiths have also given up their professions in the hope of better prospects elsewhere. Presently, the craft is practised only at a few selected places like, Nagaon, Jorhat and Barpeta to meet the limited local demand.

With a little effort this industry could easily be revived. At a time when chunky jewellery is the fashion all over the world, Assamese jewellery is bound to find a market. While retaining the essential features of the indigenous items, the designs could be modified if necessary to meet the popular requirements. But like all other items of export, it must be ensured that quality-wise and price-wise they conform to the international standard.

Industry in Packaged Food

Of late a considerable variety of indigenous food like *pitha, laru, komol chawal, sandoh guri* etc. and a number of pickles made from indigenous fruit have flooded the market. Although, at present the products cater mainly to the local market, there is immense scope for its expansion. Two enterprises that have made their mark in the national field are *Bhogali Jalpan* and *Niru's Pickles*. Starting initially as household units, they have now expanded their business outside the boundaries of North East India. It is possible that with the help of modern technology in processing and packaging, these products will find an international market.

Industry in Cashewnuts

Cashewnut has got tremendous scope in Assam. It has been successfully grown in the Garo Hills of Meghalaya and is likely to be successfully cultivated in hill slopes and upland areas of Assam, for instance, in Karbi Anglong. An integrated programme for the development of cashewnut in the region was formulated by the Indian Council of Agricultural Research and under this scheme a Regional Cashew Nursery was set up through Assam Seeds Corporation Ltd. at Ulubari, Guwahati. Further investigation is on but nothing definite has emerged. The potential is undoubtedly there. It only needs to the exploited.

RESOURCE MOBILISATION

The indigenous industries in Assam have tremendous economic potential and are likely to succeed in the present day global context. The role of the entrepreneur is extremely important because he is the one who has to manage, control and assume the risks of an enterprise. At the initial stage, however, Government support is essential. It should provide the necessary infrastructure and entrepreneurship though Government established enterprises need not be managed and owned by the Government permanently. Once the industry has been established, private enterprise should be encouraged to take over the ownership and management. In fact this had also been one of the basic principles of the colonial government. In the context of the tea industry in Assam, the Governor General and his Council were of the opinion that "it may often be the duty of the Government to act as pioneer of private enterprise"[1] and that the "Government are eventually to withdraw from the pursuance of the scheme, as soon as it may appear expedient to entrust its future success to private speculators."[2] Even an economic liberalist like J.S. Mill, while discussing the role of the State in an underdeveloped country, wrote:[3]

> "At times and places there will be no roads, docks, harbours, canals works of irrigation, hospitals,

schools, colleges, industries, unless the Government establishes them; the public being either too poor to command the necessary resources, or too little advanced in intelligence to appreciate the ends, or not sufficiently practised in joint action to be capable of the means. This is true, more or less of all countries inured to despotism and particularly of those in which there is wide distance in civilisation between the people and the Government'.

Although written in a different context and more than a hundred years ago, this is applicable to Assam even today.

Assam covers an area of 78,438 sq km. Of her total population of 25 million, around 87 per cent dwell in the rural areas.[4] The average rate of literacy in the state is around 53 per cent. Although there are both large and small scale industries in the state, Assam still remains industrially very backward. In fact, the state accounts for only 1 per cent of the total small-scale industries in the country.[5] Significantly, of the total registered small scale units in Guwahati, more than 50 per cent are non functional or closed units.[6] The situation in Guwahati is reflective of the general trend in the state.

The general problems facing this sector that we have been able to identify are:

(i) non-availability of timely finance,
(ii) frequent power failure,
(iii) unorganised production base and inadequate marketing network,

(iv) high production cost,
(v) non-availability of raw materials,
(vi) lack of skilled personnel,
(vii) stiff competition,
(viii) inadequate R&D facilities,
(ix) lack of modern technology,
(x) lack of exposure, and
(xi) government apathy. In order to enable the small and cottage industries to play their role in the development of the region, it is necessary to remove their difficulties as far as possible. Only an integrated approach to these problems can help in solving them.

Infrastructure is an economy's capital in the form of roads, railways, electric supply, water supply etc. and are facilitative structures that promote general economic activities. Unless the infrastructure is improved, an increase in production cannot be expected. The development of these sectors, however cannot be left to private enterprise because the gestation period is too long and the investments too high. Ultimately, once again it is the Government which has to step in to take over the responsibility. Assam is endowed with an infinite variety of economic resources. But much of her wealth has remained untapped because of a lack of infrastructure. Hence the region presents a paradoxical picture of being poor in the midst of plenty.

References

1. H.A. Antrobus, *A History of the Assam Company, 1839-1953*, p. 35.

2. Percival Griffiths, *The History of the Indian Tea Industry,* London, p. 53.
3. J.S. Mill, *Principles of Political Economy,* London, 1902, Book V, p. 590.
4. Census Report, 2001.
5. S.K. Saikia, *Small Scale Industries in Assam,* I.I.E. 2001, p. 4.
6. *Ibid.*

CONCLUSION

Rural industries form the backbone of a developing economy. In a state like Assam where more than 85 per cent of the people live in villages, the promotion of this sector is of utmost importance. In fact it is an urgent necessity. In the face of large-scale unemployment among the youth, industrialisation is the only way through which avenues for sustainable and productive employment can be generated. The successive Five Year Plans and Industrial Policy Resolutions have repeatedly emphasised this aspect. Apart from offering immediate large-scale employment, rural industries ensure a more equitable distribution of national income and facilitate an effective mobilisation of resources of capital and skill that would have otherwise remained untapped. Moreover, many problems that are created by unplanned urbanisation, can be avoided with the establishment of small centres of industrial production in rural areas. It is unfortunate that Assam, with her abundant natural resources, rich tradition of indigenous craft and industries and immense storehouse of knowledge in this respect, has not been able to capitalise on it. The relationship between agricultural development and industrial development is complementary rather than competitive because one is dependent on the other. A developed agricultural sector can provide a market for the manufacturers and thereby help in the growth and development of industries. In Assam especially, where the society is primarily agrarian, industries that have a link with the rural economy are likely to succeed. In this context it is important to remember that every region has its own ethos and that these ethos must be respected because no development can take place by ignoring them. Industrial development based on many of Assam's indigenous crafts can be expected to succeed primarily for this reason.

Assam does not have a long tradition of industrial investment and hence an important factor hindering industrial progress in the state is the absence of local entrepreneurs. In a developing economy, the entrepreneur is required to perform several functions. He has not only to invest the money, but also has to create the infrastructure, adopt the technology, manage the business, promote the products, and above all, undertake the risks. Apart from the absence of sufficient savings for capital investment in the rural sector, the people lack a spirit of adventure and entrepreneurship and are reluctant to undertake the tremendous risks involved in starting an industry. They are so attached to their land that they cling to it even though the holdings are often small and uneconomical. It is primarily for this reason that the industrial sector in Assam has been dominated by persons from outside the state. But the lack of capital and absence of entrepreneurs are just two aspects. A third very vital aspect is the inadequate transport facilities in the region. Difficulties in bringing the produce to the customer is often a strong deterrent in setting up an industry in the rural areas of Assam. Thus, a vicious circle of the lack of capital, absence of entrepreneurs and inadequate transport facilities have contributed to the stagnation of the economy to a large extent.

In the circumstances, we find a large exodus of skilled labour to the urban areas, where they strive to eke out a living by doing unskilled manual jobs. This trend, commonly termed as "rural skill drain", features significantly in Assam's economy today. An example in point are the several skilled workers of Samaguri Satra who, unable to make both ends meet by practising their craft in mask making, have left their homes to look for employment in the neighbouring towns.[1] Many other workers have similarly migrated to the urban areas thereby aggravating the problem of urban unemployment.

It is generally recognised that the most effective way of solving this problem is by regenerating rural industries. It is after all the small enterprises that represent the starting point and are likely to succeed in the rural sector where, in recent times, the division and sub-division of holdings have led to immense pressure on the land. The lack of capital within the economy suggests the necessity of greater reliance on labour intensive industries rather than capital intensive ones. In Assam, emphasis on agro-based industries could be a solution to the problem. At the same time attempts should be made to adopt better techniques to increase efficiency in production, lessen costs and improve quality.

Small rural enterprises can be developed independently and not in competition with the large industries. In fact, both can co-exist quite comfortably side by side. Japan has proved this more than 50 years ago.[2] In Assam, almost all the industries that we have referred to above can be left to the small-industries/cottage sector without bringing them into competition with large-scale industries. The

Government can help by creating the proper environment by removing infrastructural bottlenecks and improving transportation facilities. Government can also assist entrepreneurs at the initial stage by offering tax concessions, land at reduced rates of revenue, guaranteeing a minimum profit for a fixed period and by providing capital assistance at low rates of interest. Above all, the government should ensure peace and security in the region.

In an effort to look into these aspects, the Government has set up certain Corporations and Boards. Mention may be made ofThe Assam Industrial Development Corporation, Assam Small Industries Development Corporation, Assam Financial Corporation, Khadi and Village Industries Board, Assam Agro Industries Development Corporation, among others. Recently the Government has outlined a new approach to its industrial policy wherein it has identified its priorities to facilitate the industrialisation of the state. Development of infrastructure, improvement of communications, establishment of District Industries Centres (D.I.C.) and emphasis on the handloom and handicraft sector appear high on the list. Assam Industrial Development Corporation has been allotted government land for the development of industrial estates in different places. The D.I.C.s are expected to play a vital role in the development of small scale enterprises in the rural areas. The *Udyog Sahayak* Cell in each D.I.C. is expected to guide entrepreneurs in the matter of setting up industries. But however well meaning government policies might be, they are of no use unless they are implemented effectively. In Assam there are at present more than 25,000 registered small-scale industries, but over 45 per cent of these are either defunct or untraceable.

A major component of any enterprise is the investment of capital. In Assam there are more than 1200 branches of the different nationalised banks. In addition, there are, among others, branches of the Industrial Development Bank, Small industries Development Bank (SIDBI) and Industrial Financial Corporation of India, apart from the North Eastern Financial Corporation (NEDF) which was established a few years ago solely for financing the industrial projects of the North East. It is disheartening to note that most rural entrepreneurs are unaware of the various credit facilities and promotional schemes offered by these financial institutions. Many of those that are aware, are reluctant to avail of loans because of the unavailability of timely credit facilities, the tedious process involved in getting a loan, the complicated loan application forms, too much insistence on collateral securities and unnecessary harassment by bank officials. Indeed, all very discouraging factors. The Government, in collaboration with the banks and funding agencies, should ensure that the process for obtaining credit facilities is streamlined.

Another aspect that needs to be addressed is the inadequate Research and Development in certain sectors. We have observed, for instance, how research in *Muga* and *Eri* is woefully lacking. Moreover, it is important that the actual requirement of the rural producer be kept in mind for, in a number of cases, the link between the technology user and R&D institute is very weak.

In this context, it is perhaps not inappropriate to mention that the U.G.C. has laid special emphasis on the vocationalisation of higher education. In India, only 5 per cent of the total labour has any vocational skills in comparison to developed countries where the rate varies between 60 per cent and 80 per cent. The growing problem of unemployment has raised serious concern and it is felt that the problem can be overcome to a large extent if many opt for a skill-based education. The importance of linking education with productivity has been a major rationale for the vocationalisation of education. In Assam, keeping her natural resources in mind, the introduction of courses in sericulture, floriculture and food processing, for instance, may be considered. It is hoped that this will create a class of educated and skilled entrepreneurs primarily in the rural areas.

The competitiveness of the products of the small industrial units, both in the national and international market, is dependent to a large extent on their productivity levels, price factor and quality. At present many of the industrial products of rural Assam are not of international quality. Moreover, productivity is so limited that they are unable to meet big orders timely. If they are to survive in today's competitive world, the present products must be improved and a regular and cost effective supply ensured.

An industrial resurgence in rural Assam is the need of the hour. But development plans for industry cannot be framed in isolation. They must be integrated with plans for development of agriculture, transport, power, and all other sectors of the economy. Only then will the state progress.

References

1. S.K. Saikia, *Small Scale Industries in Assam,* I.I.E. 2001, p. 4.
2. P.C. Goswami, *The Economic Development of Assam,* New Delhi, 1988, p. 193.

INDEX

INDEX